The following is a series of photographs, many attributed to and the rest almost certainly taken by the well-known New Deal–era documentary photographer Arthur Rothstein, of massive gully erosion in Stewart County, Georgia. Many are scenes from what would become Providence Canyon State Park, the subject of this book. The Library of Congress holds these photographs, and many of them are titled, simply, "Erosion, Stewart County, Georgia." Courtesy of the Farm Security Administration / Office of War Information Photograph Collection, Library of Congress.

Let Us Now Praise Famous Gullies

environmental
history and the
american
south

SERIES EDITOR

James C. Giesen, Mississippi State University

ADVISORY BOARD

Judith Carney, University of California–Los Angeles

S. Max Edelson, University of Virginia

Robbie Ethridge, University of Mississippi

Ari Kelman, University of California–Davis

Shepard Krech III, Brown University

Megan Kate Nelson, www.historista.com

Tim Silver, Appalachian State University

Mart Stewart, Western Washington University

Paul S. Sutter, founding editor, University of Colorado, Boulder

Let Us Now Praise Famous Gullies
Providence Canyon and the Soils of the South

Paul S. Sutter

The University of Georgia Press
Athens and London

A Wormsloe
FOUNDATION
nature book

Paperback edition, 2018
© 2015 by the University of Georgia Press
Athens, Georgia 30602
www.ugapress.org
All rights reserved
Set in Adobe Garamond by Graphic Composition, Inc.,
Bogart, Georgia

Most University of Georgia Press titles are
available from popular e-book vendors.

Printed digitally

The Library of Congress has cataloged the
hardcover edition of this book as follows:
Sutter, Paul, author.
Let us now praise famous gullies : Providence Canyon
and the soils of the South / Paul S. Sutter.
xix, 248 pages : illustrations, maps ; 24 cm.
Includes bibliographical references and index.
ISBN 978-0-8203-3401-1 (hardcover : alk. paper) —
ISBN 978-0-8203-4809-4 (ebook)
1. Natural history—Georgia—Providence Canyons.
2. Tourism—Georgia—Providence Canyons—History.
3. Providence Canyons (Ga.)—History.   4. Providence
Canyons (Ga.)—Environmental conditions.   I. Title.
F292.S8 S88 2015
975.8'922—dc23

2015949031

Paperback ISBN 978-0-8203-5382-1

CONTENTS

List of Illustrations   vii

Foreword, by James C. Giesen   ix

Acknowledgments   xiii

A Note on Terminology   xix

INTRODUCTION. The Great Cut across the Face of Nature   1

PART ONE. *Arriving at Providence Canyon*   11

ONE. Yawning, Abysmal Gullies   13

TWO. The Most Picturesque Features of the Coastal Plain: Geologists Arrive at Providence Canyon   24

THREE. Rough, Gullied Land: Soil Scientists Arrive at Providence Canyon   37

PART TWO. *Making Providence Canyon Meaningful in the 1930s*   63

FOUR. A Land That Nature Built for Tourists   65

FIVE. Giving Fame and Focus to the Fact of Soil Erosion   83

PART THREE. *Returning to Providence Canyon*   109

SIX. Gullies and What They Mean   111

SEVEN. Somewhere between the Grand Canyon and a Sickening Void   158

EPILOGUE. The Ecology of Erasure   185

Notes   209

Index   241

ILLUSTRATIONS

1. U-Haul "Venture Across America" graphic featuring Providence Canyon 3
2. Physiographic regions of Georgia 15
3. The gullies of Stewart County, Georgia 18
4. "Providence Caves" in 1893 23
5. Woodcut of Lyell Gully 30
6. Frontispiece from *Second Report on the Clay Deposits of Georgia* 34
7. Milton Whitney with a National Cooperative Soil Survey map 42
8. Hugh Hammond Bennett 50
9. Colorized postcard of "Providence Canyons" 72
10. Thomas Jefferson Flanagan 78
11. "General Distribution of Erosion" 88
12. Gullies, Stewart County, Georgia, from *Soil Conservation* 91
13. Frontispiece and title page to Stuart Chase's *Rich Land, Poor Land* 100
14. Average depth of soil erosion on the Piedmont 113
15. Walker Evans, "Erosion Near Oxford Mississippi," March 1936 116
16. Walker Evans, "Erosion Near Jackson, Mississippi" 117
17. Scull Shoals ruins 120
18. Jack Delano, "Greene County, Georgia" 137
19. Jack Delano, "Erosion on a Farm West of Greensboro, Greene County, Georgia" 138
20. The Ultisols of North America 160
21. Line of cattle tick infestation in 1906 164
22. Rainfall erosivity map of the eastern United States 168
23. Arthur Rothstein, untitled, Clayton formation atop Providence sands 173
24. Arthur Rothstein, "Erosion. Stewart County, Georgia" 175
25. Gully development as a result of road drainage 178
26. Gully erosion in the Driftless Area of Wisconsin 180
27. Gully erosion on the Navajo Reservation 182

28. Tupelo Swamp, Alcovy Conservation Center   186
29. The ecology of erasure   193
30. Forest ranger and gully erosion on the Sumter National Forest   199
31. "Gully erosion by check dam. Macon County, Alabama. Tuskegee Project"   201
32. "Malakoff Diggings, Nevada County, California"   206
33. Erosional landscape at Malakoff Diggins State Park   206

FOREWORD

Irony has long been a theme of southern history, albeit a tricky one. Historians of the region have filled many shelves with books and articles describing the places and times when historical actors were unaware of the predicaments in which they were putting themselves. They take pains to describe the ways in which southerners, from the colonial era through the twentieth century, misunderstood their world. Scholars and readers alike seem to relish that surprising moment when their characters wake to find that the world they have made is not the one that they thought they were building. Appreciating unexpected and seemingly contradictory political movements, social crusades, or economic trends is a stock and trade of traditional southern history because, to put it bluntly, southern history has been ironic. The tricky thing about irony, however, is that its lessons are rarely satisfying. Its meaning is often hard to apply to other historic events or actors. As a tool for understanding the present, irony has had a frustratingly dull edge.

What follows is a book about an ironic place, Providence Canyon, a system of beautiful, historic, cavernous erosion gullies in west Georgia. Of all the remarkable things about the canyon, the irony of its creation and preservation is one of the most important. To begin to understand it, turn back to the cover for a moment, or flip again through the plates that open the volume, and look once more at the images. The striated towers of subsoil jutting up, or falling off, from the earth's surface trigger thoughts of the American West, of national parks set aside to preserve a seemingly supernatural aesthetic. But of course Providence Canyon is not the Grand Canyon. It is not an earth formation built on a millennial scale but rather the product of human work—that of a few generations rather than millions of years—on soil that once existed where now there is only air. As Paul S. Sutter explains, there is irony in a park that preserves the natural results of human-induced soil erosion, but what makes this volume truly pathbreaking and important is that this irony is merely where the story begins.

In *Let Us Now Praise Famous Gullies*, Sutter sets out to understand how the canyon came to be, why local and state officials preserved it as a park, and how the public should think about and engage with it today. It is a story that takes us from the floor of Providence Canyon to the surrounding lands of Stewart County, Georgia, around the cotton and tobacco South, through forests and government offices, and back again to Providence Canyon. It is a history not just about farmers extracting nutrients from the soil in search of profit and the land wasting away as the result. This is a history of the understandings of southern soil itself, the emergence of a science to research soil formation and weathering, and the bureaucracies that resulted. Sutter demonstrates how farmers, local officials, geologists, politicians, and a host of others made meaning first in farmland, then in the growing gullies of Stewart County. The meaning of Providence Canyon is not something fixed by the irony of its man-made origins and natural beauty, however. In fact, as Sutter explains, the meaning of Providence Canyon has never been static. Local farmers had no interest in making the growing fissure anything more than a regional oddity—in the nineteenth century they did not think much about it at all. As time passed, more and more people attempted to make something out of the gullies, to derive a lesson in the origins and existence—and yes, the irony—of Providence Canyon. Sutter reminds us as well that forgetting is itself a way to make meaning. In the final part of this book, he details the dangers of erasure, from both environmental and cultural forces, of the canyon's history.

To be sure, this book's publication is its own effort to make meaning from the gullies of west Georgia. With few exceptions, southern historians who have paid attention to the soil at all have made it out to be simply a setting, rather than an actor in the region's history. *Let Us Now Praise Famous Gullies* is important because not only has Sutter done the work to uncover the history of what farmers understood about their actions on the land, but he has shown how linked these southerners were to scientific understandings of the earth under their feet and to cultural understandings of environmental change. This shifting between the perspectives of culture and nature makes the book's conclusions not just important to historians but to anyone who thinks about the questions facing the environment of the South today. What are we preserving when we preserve the canyon as a park? Is it a monument to poor farming practices, as the displays at the welcome center tell visitors? How might a fuller history of the canyon's creation make it a more vital part of the region's ongoing conversations about environmental change? This book both raises important questions about the utility of environmental history and offers many answers.

Let Us Now Praise Famous Gullies is an important addition to the Environmental History and the American South series for many reasons beyond the merits of its analysis and argument. Sutter served as the series editor from its inception in 2007 until 2014. Now as an author, he joins those whose works he recruited and edited. Together, their scholarship has had remarkable success in shaping conversations and debates in southern history and environmental history, but most important, it has been a major factor in the creation of what is now the standalone subfield of southern environmental history. It is appropriate, therefore, that this important book by the founding editor of the series should constitute its first volume after his departure as editor. It is a book that makes clear that the Environmental History and the American South series is more than alive and well. It is poised for a bright future.

James C. Giesen

ACKNOWLEDGMENTS

Since this is a place-based study about the relationship between soil erosion and conservation in the American South, I first want to acknowledge two modest natural areas that helped to inspire this project: the Ivy Creek Natural Area outside of Charlottesville, Virginia, and the Sandy Creek Nature Center in Athens, Georgia. Both are places where my family and I would wander on weekends when we lived in these two great southern college towns. They are small natural refuges protected from residential and commercial sprawl, and they are vital centers for local environmental education. But as I came to appreciate after repeated visits, they are also landscapes with substantial and fascinating human histories, and they taught me the pleasures of reading the southern agricultural past—and particularly its erosive past—in naturalized landscapes of the present.

The Ivy Creek Natural Area, a 215-acre preserve on the western outskirts of Charlottesville, had its origin in the 1970s, when local environmental activists organized to oppose a major real estate development slated for the land. In that sense, Ivy Creek is a typical product of the rise of postwar American environmentalism. But it was decidedly *not* a pristine island in a rising sea of suburban sprawl. The Carr and Greer families, prominent African Americans in the area, had owned the land since 1870 when Hugh Carr, a former slave and sharecropper, managed to purchase it. Carr and his wife raised six children on what was then known as River View Farm. One of Carr's daughters, Mary, and her husband, Conly Greer, acquired the farm in 1914, and they farmed it for half a century while Conly Greer worked for the Virginia Agricultural Extension Division. As Albemarle County's first black extension agent, Conly worked with local African American farmers to improve their farming techniques, and he made River View into something of a model farm during his tenure. After Conly died in 1956, Mary tried to sustain River View as a working farm, but she had to sell several parcels, and other parts of it reverted to forest. Mary died in 1973, and the proposal to develop the land came soon thereafter. Today, there are several obvious markers of the land's rich human history—the family farmhouse and barn remain, there is still a family cemetery on the property, and Ivy Creek Natural Area is now an official site on the Virginia African American Heritage Trail. But as River View Farm reverted to forest through processes of ecological succession, other legacies of its agricultural past became tougher to see. As I learned more about this history, my trips to the site became less about retreating to its "nature" and more about reading the landscape as an artifact of past land use legacies. Among those legacies, despite the family's best efforts, were a series of large hillside gullies filled with old-field pines. Ivy Creek was where I first learned to read the southern landscape for its history of erosion.

When we moved to Athens, Georgia, we sought out a similar place to take our weekend walks, and we found it at the Sandy Creek Nature Center. Like Ivy Creek, Sandy Creek was founded in the 1970s and consists of several hundred acres of diverse wooded habitat, with trails that connect the nature center both to Athens downstream and to Sandy Creek Park further upstream. Sandy Creek Nature Center also had a diverse history of human land use before nature reclaimed it. There are remnants of an old brick-making factory and several ponds that had been clay pits to feed the process. There are also upland areas where past farming produced erosion and where remnant gullies are still visible. But only after getting into the research for this book did I learn that the larger Sandy Creek watershed had a central place in the history of southern soil conservation. In 1933, the newly created Soil Erosion Service (which became the Soil Conservation Service in 1935) created the Sandy Creek soil conservation demonstration project, one of dozens of demonstration areas created in 1933 and the first such project to be undertaken in Georgia. The Sandy Creek project encompassed more than one hundred thousand acres of badly eroded and gullied cotton and corn land that spanned three counties on the Georgia Piedmont. In fact, Sandy Creek itself had become so inundated with erosional debris by the early 1900s that its abandoned bottomlands regularly experienced destructive flooding, and dams along Sandy Creek—and the Oconee River into which it flowed—had silted up to the point of uselessness. Its very name, "Sandy" Creek, may have been a homage to all of this erosional debris. Like the Coon Valley in Wisconsin, which was home to the nation's first such soil conservation demonstration project, Georgia's Sandy Creek watershed was one of the cradles of American soil conservation.

Neither Ivy Creek nor Sandy Creek appear in the book that follows, but this is nonetheless a book about places like these, protected natural areas that at once conceal and reveal the anthropogenic erosion that has been a signature feature of southern environmental history. This project had its beginnings in my revelations about these places, and as you will see, it comes to rest on the importance of such places to the future of conservation.

It is a great source of pride to me that this book appears with the University of Georgia Press as part of the Environmental History and the American South series that I helped to found a decade ago. I want to thank Nicole

Mitchell, who saw the need for such a series, and the several editors with whom I worked on both the series and this book: Andrew Berzanskis, Derek Krissoff, Regan Huff, and Mick Gusinde-Duffy. As soon as I recognized that this quirky side project was becoming a book, I knew I wanted it to be in this series. I am doubly pleased that the book appears as the first volume under my successor as series editor, Jim Giesen. Jim is a superb southern and agricultural historian who first encountered environmental history as a skeptic. I would like to think that I played a small role in getting him to add "environmental historian" to his scholarly identity, but I also continue to value his critical approach to the field. His comments on the manuscript were invaluable. I also want to thank several others at University of Georgia Press: Beth Snead, who helped to oversee the editorial process; Jon Davies, who masterfully guided the manuscript through production; Chris Dodge, my wonderfully efficient copy editor; Erin New and the other designers who make such beautiful books; and David des Jardines and Jason Bennett for the marketing work that they have done and will do. Finally, I am thrilled that this book will appear as a Wormsloe Foundation Nature Book, and I thank the Wormsloe Foundation for its support. I particularly want to express my appreciation to Craig and Diana Barrow, great champions of the press who have welcomed me into their home at Wormsloe on numerous occasions and who, along with the inspired Sarah Ross, have had the vision to make Wormsloe into an important site for the study of southern environmental history.

Several other people commented on the penultimate version of this book manuscript and forced me to impose some discipline on it. Dan Richter, a prominent soil scientist (and humanities sympathizer) at Duke University who knows a lot about the soils of the South, served as an external reviewer for the press and generously revealed his identity as part of that process. This book is better for Dan's input, and I am very glad to have made his acquaintance in the process. The comments of the second press reader, who remains anonymous, were equally valuable in sharpening and tightening my analysis. My great friend Neil Maher gave the manuscript a critical read, constantly reprimanding me for my "data dumping" tendencies and other digressions. The result may not be quite as pithy as he would have liked, but Neil's comments got me closer to that ideal. Drew Swanson also gave this manuscript a generous read. Drew was one of several extraordinary graduate students in southern environmental history that I had the pleasure to advise while I was at the University of Georgia—a group that also included Bert Way, Chris Manganiello (who assisted me with some of the research for this book), Levi Van Sant, Hayden Smith, and Tom Okie. This book is the better for having been shaped by my interactions and conversations with all of them—as well as with my other master's and PhD advisees at UGA who worked outside southern environmental history, including Kathi Nehls, Ivy Holliman Way, and Michelle McQuiston.

Over the last decade, I have been lucky enough to share the research in this book with audiences all over the country. I first presented the germ of this book at the Porter L. Fortune, Jr., Symposium at the University of Mississippi, where I was an interloper among a distinguished group of anthropologists and historians that included Mart Stewart, Tim Silver, Jack Temple Kirby, Shep Krech, Don Davis, Margaret Humphreys, and Robbie Ethridge. It was Charles Reagan Wilson's kind invitation to join these "usual suspects" in southern environmental history that first got me thinking about moving my own research in a southerly direction. Several years later, I had the pleasure of presenting this research in a very different setting, New York's Adirondack Mountains, to a huge and receptive audience. It was at that meeting that Bill Cronon first suggested I write a short book on the subject. Thanks to Bill for that inspired suggestion. I also want to thank my hosts and the audiences they assembled at the University of Georgia, Macalester College, the University of Houston, Clemson University, MIT, the University of North Carolina, Western Carolina, Troy University, the University of Oklahoma, the University of Western Ontario, the University of Kansas, the University of Virginia, the University of California at Davis, and Mississippi State University.

Several people in Stewart County, Georgia, home to Providence Canyon, were critical to the completion of this book. First and foremost, I want to thank Matthew "Mac" Moye, a continuous source of information and enthusiasm. Mac not only gave me the lay of the land but also fed me a steady stream of newspaper stories from his own reading of local newspapers from the nineteenth century. Mac also connected me with several other important people in Stewart County. Sam Singer shared his memories of Providence Canyon and a few photographs as well, and I am deeply appreciative for his contributions. As I briefly note in the book itself, Bobby Williams spent the better part of a day touring me around Stewart County's back roads and showing me the county's many other giant gullies. Madge Rutledge shared her family's history with Providence Canyon and gave me some clippings and other valuable materials. Finally, Joy Joyner and the staff at Providence Canyon State Park generously allowed me to use their historical materials. One of my biggest regrets is that finishing this book at a considerable distance from Stewart County has meant that I have not been able to spend as much time there as I would have hoped.

This book is inseparable from the happy years I spent in Athens as a member of the History Department at the University of Georgia, and I want to thank my history colleagues there—and particularly the talented crowd of southern historians who tolerated my environmental carpetbagging. Several UGA scientists and Georgia friends also shaped this project in one way or another, whether they remember it or not, including Daniel Markewitz, David Leigh, Todd Rasmussen, Rhett Jackson, and J. P. Schmidt. I owe a special thanks to Peggy Galis, a friend and supporter who often provided me with a comfortable bed when I was in town and was always available for lunch at the National. I want to thank Jerry McCollum and the other Georgia Wildlife Foundation staff who welcomed me (and fed me barbecue!) when I paid a visit to the Alcovy Conservation Center. I am also indebted to Frank Magilligan, James Hyatt, and Dorothy Merritts, who willingly subjected themselves to my many questions about their research.

All historical scholarship is built on the work of others, but I feel a particular need to thank the many people whose excellent work in southern environmental history and related fields has shaped this project. I have leaned heavily in places on the work of Stanley Trimble, a pioneer in the study of southern soil erosion. Steve Stoll's work influenced how I have thought about soils and southern history. Tim Silver and Mart Stewart have been important mentors and friends as I moved into southern environmental history. I want to express special appreciation for the late Jack Temple Kirby, a true gentleman whose influence is all over this book. Southern environmental history is now a humming enterprise, and I have intellectual and personal debts to many, many others whose work has shaped this book. You know who you are, and footnotes are not enough to express my appreciation for your work. As I expanded a quirky conference paper into a book about soil erosion and soil conservation, I could not escape the suspicion that somehow Donald Worster was responsible for this digression into the dirt. This history intersects with his pioneering work on the Dust Bowl in all sorts of ways, and no one has had as great an influence on my work as he has.

Like a catfish out of muddy water, I finished this southern history back in the West, the region that first attracted me to the study of American environmental history. My new colleagues in the History Department at the University of Colorado, Boulder, made my landing here a soft one. Susan Kent was an enthusiastic supporter as department chair during my first five years here, and my new coconspirators Thomas Andrews, Lil Fenn, Patty Limerick, and Phoebe Young have made CU an exciting place to be an environmental historian. When I miss the South, I have wonderful colleagues like Virginia Anderson, Peter Wood, and William Boyd to talk with. A LEAP Associate Professor Growth Grant, funded by CU's Office of Faculty Affairs, gave me a critical course reduction that allowed me to get this manuscript done at a busy time in my career, and a Kayden Research Grant from CU's College of Arts and Sciences helped to underwrite the costs of the maps and photographs in this book.

Portions of this book have appeared in print before. I wrote a short gallery essay ("On 'Georgia's Little Grand Canyon'") about Arthur Rothstein's photos of Providence Canyon that appeared in *Environmental History* 11, no. 4 (October 2006): 830–34. I want to thank Kathy Morse for facilitating that. I also published a longer essay ("What Gullies Mean: Georgia's 'Little Grand Canyon' and Southern Environmental History") in the *Journal of Southern History* 76, no. 3 (August 2010): 579–616, that profited greatly from the editorial guidance of John Boles and Randal Hall, as well as from the comments of the journal's external reviewers.

I wandered the trails of the places that inspired this book with the loves of my life: my wife, Julie Rothschild, and our two boys, Henry and Wyatt. In fact, when I think back to our first trips to the Ivy Creek Natural Area and the Sandy Creek Nature Center, I realize that I was almost always carrying either Henry or Wyatt on my back. So it is chastening to realize that, by the time this book appears, they will be eighteen and fourteen, and Julie and I will have been married for twenty years. May our walks together continue to both inspire me and slow me down.

A NOTE ON TERMINOLOGY

Throughout this book, I refer to the region under consideration as, alternately, the tobacco and cotton South, the plantation South, and, borrowing a phrase from the geographer Charles Aiken, the plantation crescent. These are ragged descriptors, not perfect synonyms, and none quite captures the erosive South with precision. My focus on tobacco and cotton is meant to exclude southern plantation regions that produced crops such as rice and sugar, regions where plantation forms of production dominated but did not experience the substantial soil erosion that other parts of the region did. When I use "plantation South" or "plantation crescent," I am excluding those regions. My use of "plantation" might also mislead readers on several counts. First, when I use that term, I do not mean to suggest that all tobacco and cotton production, and the erosion that resulted, occurred on plantation-sized units. Rather, I mean to indicate areas where plantations were an important part, at one point or another, of the production of these crops, even if smaller farm units were also producing these staples. I often use the phrase "planters and farmers" to indicate this spectrum of productive modes. Second, my use of "plantation" is not restricted to the slave-based plantation of the antebellum era. The plantation as an economic unit outlasted slavery and marked the region under the tenant system in the late nineteenth and early twentieth centuries, even though that system involved a radical geographical reorganization of production. Some would argue that the plantation South persists to this day, although when discussing these parts of the South during the post–World War II years I refer to them as the "former plantation South." Third, while soil erosion was a problem in many parts of Appalachia, I have excluded that region from my discussion, largely because plantation production did not characterize Appalachian farming and because commentators conceptualized the environmental problems of Appalachia as discrete from those of the plantation regions. Finally, I use these terms to emphasize an important point for those new to the South and its agricultural history: there were huge portions of the region that were not devoted to tobacco and cotton (or, for that matter, sugar and rice). Aiken's notion of the plantation crescent is particularly useful for pointing to the bounded extent of these crop areas, the agricultural systems that produced them, and the soil erosion that often resulted. While these terms allow me to crudely generalize, then, the agricultural South was never a monolith.[1]

Let Us Now Praise Famous Gullies

INTRODUCTION

The Great Cut across the Face of Nature

ON ITS SURFACE, what little of it is left, Providence Canyon State Park could be the nation's most ironic conservation area. The Georgia State Parks Department describes Providence Canyon, which has long been known as Georgia's "Little Grand Canyon," as a place where "visitors are amazed by the breath-taking colors" of the canyon walls. "The pink, orange, red and purple hues of the soft canyon soil," they write in their promotional literature, "make a beautiful natural painting at this unique park." Anyone who has seen Providence Canyon would find it difficult not to agree. To the eye, it is a spectacular place, similar to the badlands and canyon country of the West except for its scale and framing vegetation. One might even call it sublime. Indeed, Providence Canyon fits neatly into the aesthetic conventions that have guided park-making in the United States for the last century and a half. Anointing it Georgia's "Little Grand Canyon" was obviously a gesture in that direction. But here's where the irony seeps in, destabilizing things: Providence Canyon is a decidedly human artifact of recent origin. The park's website admits that its gaping chasms, or gullies, which reach more than 150 feet deep and several hundred yards wide in places, were "caused by erosion due to poor farming practices during the 1800s." A placard in the visitor center instructs that "Providence Canyon is the result of man-made changes in the environment," and the center's introductory video pointedly calls the canyon "a spectacular testimony to man and his mistakes." The canyon may be home to the rare plum-leaf azalea, a source of State Parks Department pride. Wildlife may abound in the park, as the video boasts, and the gullies may "make some of the prettiest photographs in the state."[1] But Providence Canyon State Park is, by almost anyone's definition, a severely degraded landscape. Ellis Arnall, a former Georgia governor, once aptly described Providence Canyon as "the great cut across the face of nature," and the visitor center interpretation seems to agree.[2] So what does it mean to preserve and celebrate a place that is the scenic product of what its own custodians suggest were poor land-use practices? What does it mean to naturalize such a seemingly unnatural place?

{1}

Today, few people have heard of Providence Canyon or the state park that contains it. It is one of the lesser-visited parks in Georgia, and attendance is in decline. As I write, the visitor center at Providence Canyon is open only on weekends, and then only seasonally, due to state budget cuts. In fact, Providence Canyon has been demoted to a "State Outdoor Recreation Area," a moniker meant to indicate its reduced visitor services. (I will persist in calling it a state park, both for simplicity's sake and in the fervent hope that it will be restored to state park status).[3] When I asked my students at the University of Georgia, where I taught for almost a decade, if they had heard of Providence Canyon, only a few would nod in assent, and they were usually exhuming decaying memories of an elementary school field trip to the out-of-the-way spectacle. Despite its status as one of Georgia's "Seven Natural Wonders," it is today a remarkably anonymous spectacle, even as park officials and local boosters hold on to whatever shreds of evidence there is to the contrary.[4] Several years ago, for instance, the big news for advocates of the park was that U-Haul had featured Providence Canyon in one of the "Venture Across America" graphics that are plastered on the sides of their trucks and trailers (figure 1).[5] Despite such exposure, Providence Canyon is today struggling to find an audience. As it slides into obscurity, few are aware of what I discovered while researching this book: that in the 1930s, as Americans became both acutely conscious of the perils of soil erosion and eager to create national and state parks to provide for the nation's burgeoning demand for outdoor recreation, Providence Canyon was a famous place. It deserves to be famous again.

I first encountered Providence Canyon virtually, while exploring the Georgia State Parks website more than a decade ago. As an environmental historian then new to Georgia, I was eager to see what the state had to offer by way of preserved natural areas. I had some sense of the mountains of northern Georgia, the Okefenokee Swamp, and the state's well-protected barrier islands, but these areas constituted only a small portion of a large state, and I craved a broader sampling of its natural diversity. When I navigated my way to the Providence Canyon State Park site and saw a color picture of its kaleidoscopic canyon walls and hoodoo-like formations, I was stunned. I had expected this "canyon" to be in the mountains, a companion to the state's better-known Tallulah Gorge, but it is on the Coastal Plain. How could there be a canyon on the flat Coastal Plain? Then, when I read about the poor farming practices that had allegedly created the canyon, I found it hard to get past the irony of the place: the seeming incongruity and even the dark humor in a park that preserves the scenic results of an environmental disaster. As a northerner new to the Deep South, I also found myself thinking, "how southern," by which I meant how exemplary of the South's renowned boosterism to create a park

Figure 1. U-Haul "Venture Across America" graphic featuring Providence Canyon, with a flowering plum-leaf azalea in the foreground. Courtesy of U-Haul International.

from such an ignominious example of the region's land use history. It seemed to me a particularly southern gesture to turn a scar into a point of pride.

I first visited Providence Canyon a couple of years later. I was traveling with my family to the Florida Gulf Coast for a beach vacation, and I insisted on the kind of environmental history detour that they have gotten used to enduring. We pulled off Interstate 75 southbound at Vienna and headed west to Americus. We stopped in Plains, Jimmy Carter's hometown, to eat lunch and poke around its small downtown. It was late October, so the worst of the humidity had dissipated for the year, but it was still warm, and there was a shimmer to the air. After piling back into the car, we headed west, through the towns of Richland and Lumpkin, until we reached the park's entrance gate. The land around Providence Canyon was nothing special. Mostly it was planted pine or scrubby old-field succession of the sort that one who lives in the South is well used to seeing. A loblolly stand near the park entrance had just been harvested, and slash piles remained along the edge of the road. It was not the gateway scene one might expect at the entrance to a park, but this was no regular park. As we left the car and approached the fence line that guarded the canyon's rim, our jaws slackened. The various fingers of the canyon were deeper than I had expected, and the colors of the canyon walls were indeed striking in their variety. Providence Canyon is nothing comparable to the Grand Canyon proper, of course, but it did remind us of other western parkscapes such as Utah's Bryce Canyon or South Dakota's Badlands. My wife began peppering me with questions about the causes of the erosion to which we were witness. I answered that it had been the result of cotton farming some time in the nineteenth century. She laughed, as many people do. Then we got back in the car and drove to meet the friends with whom we were sharing a beach house. That night, over dinner, we recounted our brief stop and shared some photographs. We agreed that it was a bizarre phenomenon and left it at that. But that visit stuck with me, and when a conference invitation soon thereafter gave me the opportunity to write on a southern environmental history topic of my choice, I decided to take a stab at making sense of the place. I assumed that my essay would be a playful exercise, a chance for a light but clever meditation on the irony of these big gullies that had been preserved as a park. But I soon discovered that there was more to the history of the place than I could have imagined, more gravity to accompany the levity.

Most important, I realized that focusing on irony did Providence Canyon a disservice. There was something deeper there that irony could not plumb. I came to understand that the logic behind my immediate sense of irony was faulty, for it presumed a strict policing of the boundaries between the natural and the unnatural that did not map onto the history of the place. Irony was getting in the way of my seeing Providence Canyon clearly. As for its southernness, I sensed that that too was up for debate. I remain convinced that there is much that is distinctly southern about Providence Canyon and that one can learn a lot about the soils of the South and the region's larger environmental history by staring into its multihued abyss. A large portion of this book will be devoted to making that point. But Providence Canyon should not to be seen as only a southern site, for it can teach us both smaller and larger lessons about conservation and environmental history. This book, then, is my effort to figure out what Providence Canyon has meant and should mean as a natural and cultural landscape—and as a landscape of conservation. As we will see, I am not the first person to make such an effort.

I borrow my title from an iconoclastic Depression-era study of sharecroppers in Hale County, Alabama. In 1936, *Fortune* magazine, one of the most influential outlets for documentary journalism during the Great Depression, sent the writer James Agee and the photographer Walker Evans to the Deep South to document the conditions of white sharecroppers. The result was supposed to be a feature article by Agee, illustrated by Evans's photographs, that would stir the consciences of *Fortune*'s business-minded readership. The Great Depression brought considerable documentary attention to the plight of southern tenant farmers, particularly the poor white farmers whose numbers had swelled under the crop-lien system. But Agee and Evans chafed at the assignment, especially at how *Fortune*'s editors asked them to objectify and render as typical the lives of three families just barely scraping by. By reducing these families' lives to mere exhibits of the impoverished conditions many experienced in the tenant-farming South, their individuality and dignity, Evans and Agee feared, would be swallowed by the magazine's representational goals. The camera, and the writing commissioned to accompany its images, threatened to steal the families' souls.[6]

Famously, Agee delivered an article manuscript, now lost, that was at least ten times longer than *Fortune*'s editor had asked for and that was too detailed and painfully subjective to pass as conventional journalism. The critic Dwight Macdonald described it as "pessimistic, unconstructive, impractical, indignant, lyrical, and always personal."[7] After sitting on the manuscript for a year, *Fortune* rejected it. Over the next two years, Agee transformed his lengthy article into a book, but even then the interested publisher, Harper and Brothers, demanded extensive revisions that bent the piece back in the direction of conventional documentary journalism. Agee refused, and so Harper and Brothers rejected the manuscript too. Finally, when Agee was ready to give up on the project, the manuscript fortuitously fell into the hands of Paul Brooks, Houghton Mifflin's legendary editor in chief who would later become well known, at least in environmental circles, as Rachel Carson's editor and good friend. Brooks accepted the manuscript with only minor revisions, mostly requesting that it be expunged of profanity. The result, *Let Us Now Praise Famous Men*, appeared in 1941, just as World War II was undoing any lingering interest American readers might have had in the lives and struggles of southern tenant farmers.[8]

Despite poor timing and lackluster sales, *Let Us Now Praise Famous Men* steadily rose in the estimation of critics. The book began with a series of Walker Evans's photographs that mixed sympathetic portraiture with the bare facts of these families' existences. Evans's images captured the lives of the Gudger, Burroughs, and Ricketts families as both mundane and transcendent. Agee's text then alternately sketched the sharecroppers' world in excruciating detail and banged against the prison bars of his own subjectivity. The brilliant but troubling book that emerged from their inquiry stands today as the most powerful critique of the limitations of the New Deal documentary impulse.

Evans and Agee refused to simplify and package, for propagandistic ends, the complex lives of the poor and powerless in word and image. They refused to allow these three families to stand for a larger set of social problems.

My aim in *Let Us Now Praise Famous Gullies* is a parallel one, and I pay homage to Evans and Agee not only by invoking their title but also by including a section of photographs—in this case by Arthur Rothstein, another well-known New Deal–era documentary photographer—at the beginning of the book.[9] But this is a different book than Evans and Agee's. I contend primarily with the environmental rather than the social politics of representation and reform, and with the consequences of objectifying landscapes rather than individuals. I argue that different groups of people historically represented, and in the process oversimplified, the nature and history of Providence Canyon to serve their own ends. Like Evans and Agee, I want readers to think critically about what it means to use an object, in this case a badly eroded landscape, as an emblem of a larger set of historical processes. I want readers to begin with Rothstein's images, see what sorts of moral stories well up inside of them, and then follow me through a process of rethinking those narratives.

My story unfolds in three discrete parts. The core of the book, part 2, explores the surprising history of an interwar contest over Providence Canyon's meaning, a contest that I discovered through Rothstein's photographs. On one side of this contest was a group of New Deal soil conservationists and environmental reformers who insisted that Providence Canyon was a human-induced disaster representative of a larger southern history of soil abuse, and who used its example frequently in their writings, often with accompanying photographs. I found Providence Canyon was everywhere in the New Deal environmental literature, hiding in plain sight. On the other side of this contest were local and regional boosters who, in the late 1930s, wanted to preserve Providence Canyon as a national park, and who claimed, incredibly, that it was natural. They mounted a substantial campaign to achieve this goal, but they fell short and had to settle for state park status several decades later. This unlikely contest between those who saw Providence Canyon as an expression of nature and those who saw it as a product of culture sits at the center of this story.

In trying to make sense of this interwar contest over what Providence Canyon meant, however, I found myself asking deeper historical questions about where Providence Canyon came from and why people began paying attention to it when they did. While the exact origins of Providence Canyon are hard to determine, it was well formed before the Civil War. But it took more than half a century for anyone beyond its locale to see it as a noteworthy place. Even locals seem to have been only marginally interested in its growing grandeur during the second half of the nineteenth century. What, then, might explain its late arrival as a popular visual spectacle? Answering that question unexpectedly pulled me into an exploration of the rise of soil science, its disciplinary drift away from geology, and the ways in which that separation laid the intellectual groundwork for the divergent interpretations of Providence Canyon in the 1930s. Part 1, then, provides the deep background for the 1930s story.

Finally, in part 3, I offer my own interpretation of Providence Canyon. I take seriously the representative impulse of New Deal soil conservationists and environmental reformers, and I try to make sense of the larger history of soil erosion in the South and how Providence Canyon might be made to represent that history. In doing so, I pay close attention to both the human and the environmental factors that shaped this regional history of erosive agriculture. Again, my survey often takes me far from Providence Canyon, though I circle back to the site and rethink its place in this larger regional history of erosion. While I do see Providence Canyon as a representative place, I also argue that critical aspects of its history are sui generis, and that using Providence Canyon to argue for regional distinctiveness obscures some important lessons about the local. Finally, I attempt to dismantle my initial sense of irony by comparing Providence Canyon State Park to other landscapes of conservation that have similar histories. In the end, I conclude that seeing Providence Canyon only as a result of "poor farming practices," only as "a spectacular testimony to man and his mistakes," misses crucial aspects of its environmental history. In fact, I have come to see Providence Canyon as a more natural place than my initial sense of irony would allow.

There are several timely reasons for boosting Providence Canyon to a place of public prominence once again. Environmental historians have mapped the nature-culture divide with considerable energy in recent decades, and they have often played with the irony that places we thought were natural are often deeply shaped by human culture. This has been particularly true for studies of parks and other landscapes of conservation. Most environmental historians now recognize national parks and wilderness areas as profoundly cultural products. Not only are they physical manifestations of ideas that emerged at particular historical moments and under specific political, social, and cultural conditions; they are also landscapes that we manage so that they appear untouched, often by removing people and signs of their past activities and by employing the skills of landscape architects to smooth the rough edges of visitor infrastructure. A tremendous amount of work, intellectual and physical, has gone into making parks seem pristine, and such work has erased a lot of human and natural history in the process. This critique is a compelling one, but it can so highlight the tarnishing qualities of human artifice that we now have a hard time making sense of the natural value of these preserved places. If parks and wilderness areas are not purely natural, what are they? Are they purely cultural creations? Are they places where nature and culture intermix to create some sort of conservation alloy, what we environmental historians have come to call "hybrid" landscapes? Irony may be a compelling way of presenting this problem of culture in nature, but it is analytically limiting when it comes to resolving it.

What is refreshing about Providence Canyon State Park as a case study in this genre is how unsubtle it is. Providence Canyon is so obviously *not* pristine that using its history to trouble environmentalist assumptions is too easy an exercise. As an artifact of historical human land use, Providence Canyon cries out for analysis that moves in a countervailing direction, from the cultural to the natural. Without giving away too much here, I ultimately argue that Providence Canyon not only deserves to be a park but that it is not too different from many of our other landscapes of conservation, and that it might offer us a model for thinking about those places in ways that move us beyond the rigidity, and the easy ironies, of the nature-culture divide.

Providence Canyon also serves as a useful window on interwar conservation and New Deal environmental history. Environmental historians have thoroughly documented the history of Progressive-era conservation, and they are giving us a fuller sense of the postwar environmental movement as well. But with a few notable exceptions, they have only recently turned their attention to the interwar years.[10] This is surprising because the New Deal was one of the most formative eras in shaping our modern environmental consciousness, and it was a time in which conservation thinkers were surprisingly attentive to the nation's environmental history. If there was any singular concern that marked the New Deal era as distinct from those that came before and after, it was a concern with soil, its centrality to human civilizations, the history of its use and abuse, the rapidity with which it seemed to be wasting away, and the tragic human consequences of soil improvidence. Erosion even functioned as a powerful *social* metaphor during the Great Depression, describing the human poverty that corresponded with eroded lands. The New Deal documentary impulse focused on environmental degradation as an ill to be logically paired with poverty, though again, with a few notable exceptions, we have paid little attention to the formative influence of New Deal environmental representation and rhetoric.[11] This is a history, then, that quietly attempts to chart a central place for both soil and the New Deal in the nation's modern environmental history.

Finally, I want to use Providence Canyon to think about the environmental history of the American South. Environmental historians were slow to turn to the South and its histories of land use and conservation, though in recent years that neglect has dissipated substantially. There were several reasons for this delayed interest. Compared to other regions, the South was slow to develop a tradition of environmental activism, and to the extent that environmental history scholarship emerged from activist interest, studies of the South's environmental history have lagged as a result. Moreover, the South, as a predominantly agrarian region, also lacked the wilderness tradition and public lands politics that animated so much early American environmental history scholarship.[12] Perhaps most importantly, there is an older tradition in southern historiography of environmental determinism—of historical arguments that saw environmental forces as *the* determining causes in ways that allowed little room for human agency and historical contingency—that made postwar historians of the South understandably hesitant to explore the region's environmental history, fearing that such efforts might diminish the shaping influence of human actions and decisions. The burden of southern environmental history, then, has been to reintroduce the environment as a causal force while avoiding the determinisms of the past.[13]

Providence Canyon provides an opportunity to discuss the neglected environmental history of the tobacco and cotton South, a subregion that experienced a dramatic history of soil erosion and environmental decline. In quite a few cases the presence of vast gullies marked the passage of tobacco and cotton culture as it spread across the region, and the evidence of those ravages were at their most visible, and photogenic, during the New Deal era. This story of staple crop agriculture and the erosion that it produced ought to be central to the South's environmental history, but both southern and environmental historians alike have largely neglected it. This book cannot fully tell that story, but I hope it will provide a framework for others who will.

Let us now praise famous gullies.

PART ONE

Arriving at Providence Canyon

IT TOOK SOME TIME to arrive at Providence Canyon. I do not mean that literally, as if describing one of my many visits to the park, though Providence Canyon is not easy to get to and will need locating before I move to the heart of the story. Rather, there was a significant lag of almost half a century between the initial development of the gullies that would come to compose Providence Canyon State Park and their discovery by outside observers intent on assigning them significance. Just as Providence Canyon is today a little-known local curiosity, so it was during its early history. Chapter 1 of this book provides a brief geographical and historical introduction to Providence Canyon, though I withhold important information about how these gullies developed for the end of the story. We do not want to arrive at Providence Canyon too quickly.

In the years just before World War I, two groups of scientific surveyors began showing modest interest in these spectacular gullies, making them known to broader constituencies. The first group to take in the visual spectacle was the Geological Survey of Georgia, practical scientists who were engaged in a revived effort to identify and map the state's mineral resources. The second group to note the area's broken territory was the National Cooperative Soil Survey, a joint federal-state effort to map the nation's soil resources. Neither group paid more than glancing attention to Providence Canyon, at least initially, but the histories of each of these surveying efforts are critical to understanding the contest over Providence Canyon that developed in the 1930s. They help us make sense of why scientific attention arrived at the place at a particular moment in time, but they also speak to a shifting set of scientific and cultural understandings of geology, soils, and sublime scenery that soon made Providence Canyon a contested place. Chapter 2 examines the first part of that story: the history of geology and geological surveying in the nineteenth century and the rise of a geological aesthetic that would eventually contribute to Providence Canyon's reputation as Georgia's "Little Grand Canyon." Since the study of soils was a subset of geol-

{11}

ogy for much of this period, this chapter also contends with the early history of soil science and with the drifting apart of the two inquiries in the late nineteenth century. Chapter 3 focuses on the rise of the Soil Survey, the discovery by soil surveyors of the gullies of Stewart County, and the ways in which that discovery helped to set the stage for the emergence of soil conservation by the late 1920s.

CHAPTER ONE

Yawning, Abysmal Gullies

PROVIDENCE CANYON STATE PARK is located in Stewart County, Georgia, in the west-central part of the state, about forty miles south of the fall-line city of Columbus. Fort Benning, the home of the U.S. Army Infantry School, sits just to the north in Chattahoochee County, which separates rural Stewart County from metropolitan Muscogee County. Stewart County's western boundary is the Chattahoochee River, much of it impounded in Lake Walter F. George, named to honor one of Georgia's longtime U.S. senators. Alabama begins somewhere in the middle of the lake (Alabamans prefer to call it Lake Eufala), which is renowned for its bass, the game fish that has thrived in the novel ecological conditions of the South's many dammed rivers. When the people of Georgia first elected Jimmy Carter as a state representative in a controversial 1962 election, Stewart County was in his newly drawn district, and by the time Providence Canyon became a state park in 1971, Carter had risen to the governorship.[1] Indeed, the Carter family played an important role in Providence Canyon's eventual preservation, and their involvement is one of its claims to fame. Today Stewart County is sparsely populated, with only about six thousand residents, and it is poor, with a quarter of the population living below the poverty line, almost double the national average. Stewart County is also heavily forested, with about 90 percent of the county land base now covered with trees, many of them plantation pine.[2] But Stewart County was once one of the premier cotton-growing counties in Georgia, and in the entire South for that matter. The county's massive gullies date from that former life.

Stewart County sits on the cusp of several major southern physiographic regions and on the margins of what were once distinctive plantation areas. Physiographically, Stewart County is a part of the upper Coastal Plain. The county's northern boundary is about twenty miles to the south of the Coastal Plain's border with the Piedmont, the heart of the tobacco and cotton South in the late eighteenth and nineteenth centuries and the region better known for its long history of human-induced soil erosion. Stewart County is just to the south of the southwestern and last-settled portion of the Piedmont, and it lies squarely within the Fall Line Hills, a belt of rolling territory that extends south from the fall line and widens as it moves from northeast to southwest across the region. Geologically, the Fall Line Hills mark the meet-

{13}

ing place between the sedimentary Coastal Plain and the much older hard-rock geology of the Piedmont. During periods of higher sea levels in the deep past, advancing and retreating shorelines shaped these hills.[3] As a result, the county has unusual topographical relief for the Coastal Plain. From the Chattahoochee River, Stewart County rises through its hilly western portion to an upland plain almost seven hundred feet above sea level at its highest point. Just across the Chattahoochee from Stewart County lies the eastern edge of the Black Belt prairies, a crescent of lime-enriched and once-fertile soils that sweeps west and then north for several hundred miles through Alabama and into Mississippi. This is the physiographic Black Belt, not to be confused with the more expansive demographic Black Belt defined primarily by its plantation past and a large African American population. The Black Belt prairies, which were largely treeless in their precontact state, were rapidly settled and converted to cotton culture in the decades before the Civil War, an extension of the wave of settlement that spilled off the Piedmont and populated Stewart County. They contained some of the plantation South's most productive soils, but this region, like the Piedmont, soon showed serious signs of wear and tear. Stewart County also sits on the northwestern edge of the southwest Georgia plantation region that was centered on Albany, Georgia. Again, beginning in the 1820s, cotton planters and their slaves arrived and converted the red lime soils of the Daugherty Plain along the Flint River into an outpost of the cotton kingdom. It was a place where, according to historian Susan O'Donovan, "slave society arrived fast, ferociously, and late."[4] By the eve of the Civil War, Southwest Georgia was one of the richest and most brutal regions of the cotton South. W. E. B. Du Bois later called it the "heart of the Black Belt," a place where "the corner-stone of the Cotton Kingdom was laid." By the time of the Civil War, Du Bois noted, many also called it the "Egypt of the Confederacy."[5] Stewart County's simultaneous situation at the center of the historic "plantation crescent," to use Charles Aiken's useful descriptor, and its place on the margins of several key physiographic and cultural regions suggest some of the complexities involved in categorizing the place.[6]

Before Euro-American settlers and African American slaves began to arrive in the early nineteenth century, Stewart County had been a central place for southeastern Indian peoples for centuries, and its environment was shaped by their presence. Stewart County's river frontage was the site of significant Mississippian settlements, and the county has two major mound systems along the Chattahoochee River to prove it.[7] But those societies fell apart by the early seventeenth century as a result of Old World diseases and the political and social instability they wrought. In place of these Mississippian settlements there developed a series of "coalescent societies," new polities such as the Creeks,

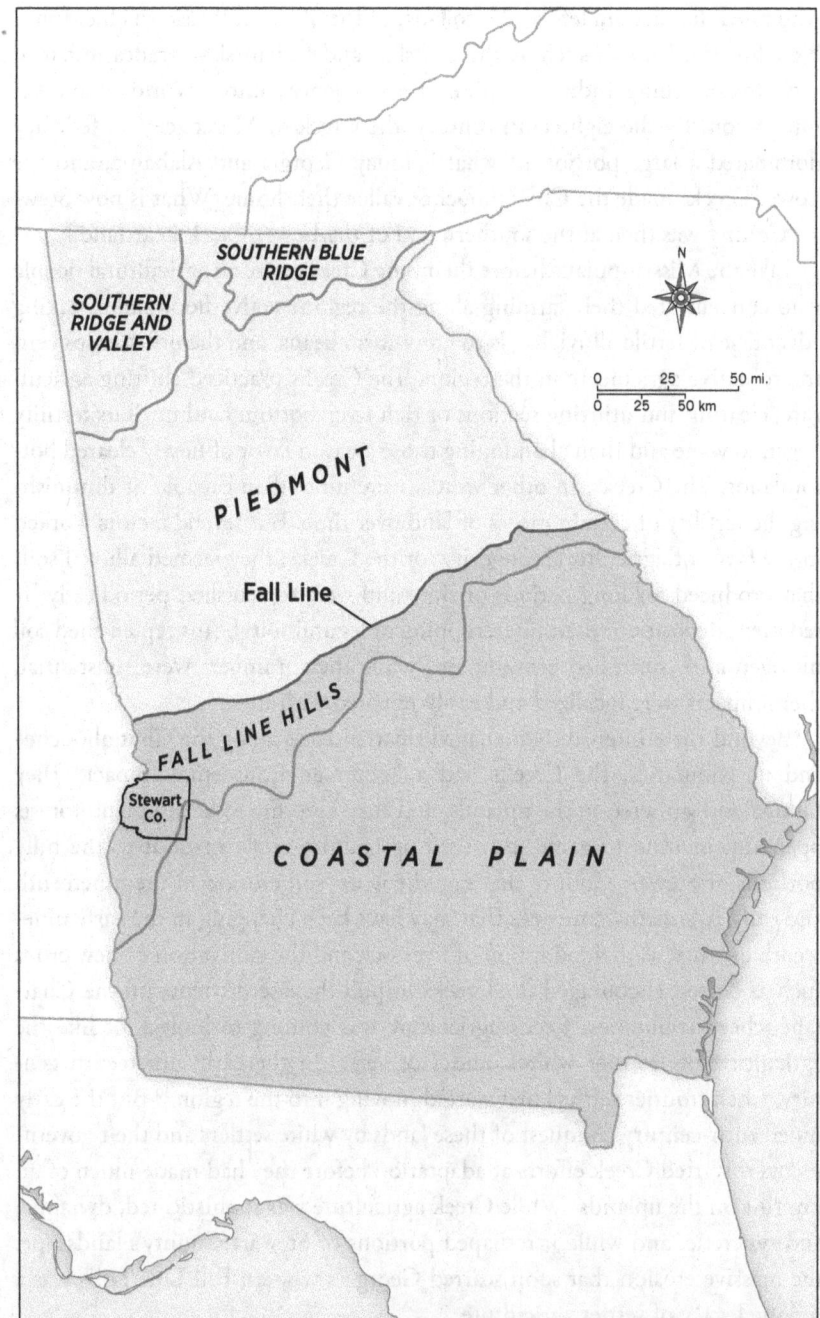

Figure 2. Physiographic regions of Georgia, adapted from Robert McVety's master's thesis. Note Stewart County's location in the west-central part of the state, within the Fall Line Hills of the upper Coastal Plain.

who filled the vacuum left by the collapse of the older southeastern chiefdoms. New historical forces such as the deerskin and Indian slave trades reshaped their lives, pulling Indian peoples, often violently, into a world of market integration. By the eighteenth century, the Creek or Muscogee Confederacy dominated a large portion of what is today Georgia and Alabama, and the Lower Creeks made the Chattahoochee valley their home. What is now Stewart County was then at the southern end of the Lower Creek heartland.[8]

Like the Mississippians before them, the Creeks were an agricultural people who concentrated their farming along the region's active floodplains, taking advantage of fertile alluvial soils to grow corn, beans, and the other crops central to native agriculture in the region. The Creeks practiced shifting agriculture, clearing and utilizing sections of rich river-bottom land until its fertility began to wane and then abandoning those areas in favor of newly cleared bottomlands. The Creeks, in other words, were more than capable of diminishing the fertility of certain pieces of land over time. But several factors worked in the favor of agricultural longevity for the Creeks: they farmed alluvial soils that produced for long periods of time and were replenished periodically by sediment deposition; their intercropping of leguminous beans replenished soil nitrogen and controlled erosion; and while their numbers were substantial, their impacts were localized and easily repaired with time.[9]

Beyond these intensively managed riparian areas along the Chattahoochee and its tributaries, the Creeks had a lighter environmental impact. They hunted and gathered in the uplands, and they used fire to keep upland forests open and inviting to game. But their agriculture rarely made it to the hilly portions of Stewart County that saw the worst soil erosion in the nineteenth and early twentieth centuries. That may have been changing in the early nineteenth century, as their adoption of livestock and the cultivation of new crops such as cotton encouraged the Creeks to pull their settlements up the Chattahoochee's tributaries. Creek agriculture was coming to look a lot like the agriculture of frontier whites, and vice versa, in the early nineteenth century, when frontier settlers first started moving into the region.[10] But the early nineteenth-century conquest of these lands by white settlers and their governments thwarted Creek efforts at adaptation before they had made much of an imprint on the uplands. While Creek agriculture was sophisticated, dynamic, and syncretic, and while it reshaped portions of Stewart County's landscape, the massive erosion that soon scarred Georgia's western Fall Line Hills was a unique legacy of settler agriculture.

The network of canyons that makes up Providence Canyon State Park—the park actually contains sixteen steep-sided gullies—covers only eleven hundred acres, but large parts of Stewart County, and portions of several adjoining counties, are riddled with similar gullies.[11] In the early twentieth century, soil surveyors estimated that forty thousand acres in the county, about 12 percent of its land base, were "rough, gullied lands." In the late 1960s and early 1970s, Robert McVety, a graduate student in the Department of Geography at Florida State University, used satellite photos and the knowledge of long-time local residents to map as many of the steep-sided gullies in the county as he could find. When his work was done, he had found 159 substantial gullies (see figure 3), and he assumed that he missed a few because of the extensive revegetation that had occurred over the previous half-century, as agriculture retreated from the county.[12]

In search of some of these other gully specimens, I asked Bobby Williams, a long-time employee of the timber company MeadWestvaco and by several accounts the local resident with the keenest on-the-ground knowledge of the county, to give me a gully tour. He graciously obliged. I met Bobby one morning at the MeadWestvaco offices several miles north of Lumpkin, and he told me that he had half a dozen sites that he wanted to show me. In his enthusiasm, he ended up introducing me to many more. Within just a few hours of hitting the county's network of unpaved secondary and timber company roads, we saw numerous prime specimens, a few of which rivaled Providence Canyon in scale and scenic qualities. Some were draped in kudzu and hard to make out, and one had been partially filled with trash before stricter sanitary landfill regulations came into effect several decades ago. But there are several enormous gullies still open to view if you know where to find them. Most of these other gullies remain little known, however. Only those that constitute Providence Canyon have become a state park and minor tourist destination. Discovering the others requires a sense of adventure, a good tour guide, and, in some cases, a machete.

The gullies of Stewart County began to form during the county's early frontier settlement, some within a generation of the arrival of the first settlers, and they yawned ever wider as cotton took hold of the region. There is general consensus on that, although, as we will see, there are some intriguing theories about just how gully erosion began in the county. With Creek claims to the region extinguished by the Treaty of Indian Springs in 1825, the State of Georgia surveyed the land and then opened it to settlement in 1827 as part of the state's fifth land lottery, though the first settlers had arrived a few years before that. The county, created from an adjoining one in 1830, was named for General Daniel Stewart, a native Georgian who served in both the Revolutionary War and the War of 1812 but whose greater claim to posthumous fame would be his great-grandson Theodore Roosevelt. Conflict between settlers and Indians in the county persisted into the mid-1830s, and it peaked in 1836

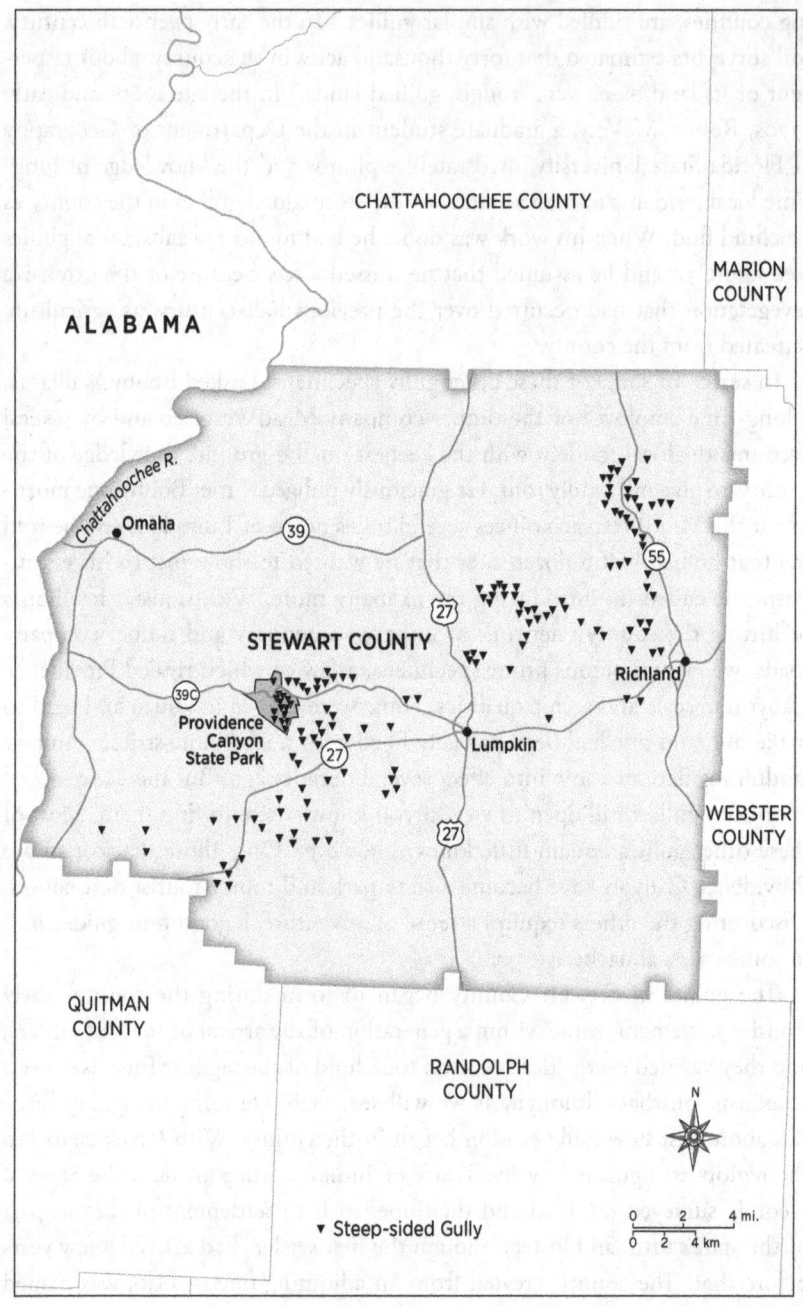

Figure 3. The gullies of Stewart County, Georgia. In a master's thesis, Robert McVety located all of the steep-sided gullies that he could find in Stewart County. This map, adapted from McVety's work, shows the extent of the county's gullying. The cluster of gullies known as Providence Canyon is due west of Lumpkin.

when a group of several hundred Creeks burned the Roanoke settlement on the Chattahoochee River to the ground. Local residents called in the state militia to put down Creek raids, and fighting effectively ended with the Battle of Shepherd's Plantation in June 1836. Settlement proceeded quickly thereafter.[13] By 1840 there were 3,741 people in the county, all but one of whom was engaged in agriculture. Only five years later, there were 14,241 county residents, 5,744 of whom were slaves. By 1850 the slave numbers had jumped to 7,373 out of a total population of 16,027. The county's population hovered around 15,000, about two and a half times its current level, for the rest of the century. As early as 1850, the county produced 7.6 million pounds of cotton, which made it one of the top three cotton-producing counties in the state.[14]

Much of the early settlement of Stewart County fanned out along the fertile floodplain of the Chattahoochee River. Like the Creeks before them, early settlers and land speculators favored bottomlands as farming sites because of their accreted fertility, as long as they were not too swampy, malarious, or difficult to clear. Wealthy plantation owners—including, most prominently, General Robert Toombs, a Georgia senator who would briefly serve as the secretary of state for the Confederacy—soon controlled most of the county's bottomlands. These lands were not only good for farming; they also placed planters just below the fall line and right along one of the most important transportation routes for marketing their crops.[15]

Following settlement patterns established on the Piedmont, planters and farmers shut out of the fertile but limited floodplain lands focused their energies on the adjoining and undulating section of the western part of the county, which was a mixed forest of oak, hickory, and shortleaf pine before agricultural clearance. Most settlers avoided the level lands in the eastern part of the county where longleaf pines dominated, following the general assumption of the era that pinelands were less fertile than those with hardwoods. This was not a bad assumption, as longleaf pines grow on acidic and well-drained lands that can be difficult to farm. There is also some evidence that the hilly sections of the county had an abundance of springs, which gave early settlers access to fresh water for themselves and whatever livestock they brought with them. Proximity to the river and the inexpensive transportation it provided was also important for those who wanted to produce cotton for distant markets, which again favored the hilly western part of the county over its flatter eastern part. But this choice to favor the rolling over the flatlands of the county was a fateful one in terms of the erosion history to come.[16]

Gully erosion came quickly to the newly settled county. Among the first settlers in the immediate vicinity of what would become Providence Canyon were the Reverend David Walker Lowe and his wife, Jane Dorsey Lowe. David

Lowe was a Methodist circuit rider from South Carolina who moved to Stewart County from Warren County, in Georgia's eastern Piedmont, in the mid-1820s. After building a house in western Stewart County, Lowe founded and became a charter member of the Providence Methodist Church, organized in 1832–33 and originally built as a rude log structure on several acres of land donated by Lowe on the south side of the Lumpkin–Florence Road. Lumpkin, then and still the county seat, was about seven miles to the east. Florence, a substantial settlement on the Chattahoochee, was about ten miles to the northwest. (The settlement of Florence no longer exists. Today Florence State Park and Marina occupies the former town site, or at least that part of it not now under Lake Walter George). Lowe and his congregants began clearing the land of its forest cover and farming and grazing the territory surrounding the church in the early 1830s. It was not long before the land began slipping away from them. There is little solid evidence about exactly when or how this happened, but we do know that, by 1859, gully erosion had become so bad in the vicinity of the church that the congregation had to move to a new house of worship on the north side of the road to avoid sliding into the expanding chasm. Several other buildings, including a barn and a schoolhouse, and many of those buried in the church graveyard were consumed by the gully. Another gully complex, which is still visible on the right side of the road as one approaches the park from the east, began to develop to the north of the church, such that, in the decades after the Civil War, the community found itself on a narrowing isthmus of high land. Lowe, who passed away in 1843, did not live to see the construction of the new church, which still sits on a parcel of private land surrounded by the state park. But he may have lived long enough to witness the incipient gullies that would later be named for the church that he founded.[17]

It is hard to find reliable sources on how the earliest settlers reacted as gullies developed throughout the county in the years before the Civil War, though Providence Canyon's name hints at what was likely the dominant interpretation of "canyon" formation: that this was an act of God, though not a particularly providential one, it must be admitted.[18] One wonders what Lowe and his successors made of the emerging spectacle in their Sunday sermonizing and how their congregants might have responded to the land giving out under their feet. Perhaps they worried that the wasting away of the land was a sign of God's displeasure with the congregation. We know that such was a common interpretation of natural disasters up through the mid-nineteenth century. As environmental historian Ted Steinberg has noted, "Seeing floods, earthquakes, and storms as signs of God's displeasure is arguably one of the oldest ways of interpreting these events."[19] However they interpreted the divine in these gullies, it is unlikely that Lowe or his congregants concluded that God was displeased with the basic goal of clearing the land to make it agriculturally productive. "Improvement" was a dominant environmental ideal in nineteenth-century America and one imbued with a religious mission. Whether the locals expressed what we would today call conservation concern as their land collapsed and washed down toward the Chattahoochee River is hard to say.

Concern for environmental degradation did develop in small pockets of the United States before the Civil War, and portions of the southern agricultural landscape stood as proto-conservationist object lessons during this period. Conventional wisdom says that it was George Perkins Marsh's landmark book of 1864, *Man and Nature; or, The Earth as Modified by Human Action*, which planted the seeds of modern conservation consciousness in the United States. Marsh had observed frontier environmental degradation firsthand in his native Vermont, and as a result of extensive travel, he came to argue that similar patterns of environmental destruction had occurred centuries ago in the Old World as well. For Marsh, soil care was one of the foundations of cultural longevity, and many a previous civilization had been buried by its erosive tendencies.[20] But well before Marsh's landmark study appeared in print, he was but one among a group of antebellum agricultural reformers, many of them southern, who critiqued wasteful farming practices, particularly the unwillingness of frontier farmers to care for and build up the soil.

These critiques of American, and particularly southern, agriculture began to appear as early as the late eighteenth century. Not long after the American Revolution, the great Virginia revolutionary Patrick Henry is said to have remarked: "Since the achievement of our independence, he is the greatest Patriot, who stops the most gullies."[21] By the early nineteenth century there were numerous observers in, and of, the tobacco and cotton South who were reaching similar conclusions. Some of these observers were northerners and European travelers who made a sport of heaping scorn on southern land use practices. But there were also native southerners worried by the exhausted and eroded lands that spread across the region. They constituted, as the historian Steven Stoll has argued, an early though neglected strand of American conservation concern.[22] These observers produced a southern variant on this discussion of land abuse that flourished in the antebellum era, but they did so mostly from seats in the settled east, from towns and cities, or from the confines of a few model Piedmont farms.[23] The city of Columbus, not far to the north of Stewart County, did briefly host an influential agricultural reform journal, *The Soil of the South*, from 1851 to 1856, but it is unclear whether the farmers of Stewart County read this journal or adopted its prescriptions.[24] Such "book farming," as many dismissively called it, did not reach far down the social ladder or out onto the frontier. Rather, it was the "wasty ways" of frontiersmen, to borrow novelist James Fenimore Cooper's apt phrase, that ruled the day in places such as antebellum Stewart County.[25] Had Patrick Henry been able to visit Stewart County on the eve of the Civil War, he might have worried about its residents' patriotism.

The gullies of Stewart County expanded unnoticed by the outside world during the final decades of the nineteenth century, until they covered tens of thousands of acres by the early twentieth century. An 1893 photograph, the earliest one that I found of the gullies that would come to be known as Providence Canyon, shows that they were huge before the end of the century (see figure 4). Local resident R. T. Humber, who may have been the owner of some of the collapsing land, hired a commercial photographer from nearby Eufala, Alabama, to make a visual record of the spectacle. This photograph includes three seated figures in the lower right—from left to right, Arthur Willett, Roena Willett, and Will Humber—whose presence provides a sense of the scale of these formations. The camera captured the place at a moment when photography was just becoming an important and widespread part of American culture and commerce. Indeed, the 1890s was the beginning of the golden age of the commercial photographer in American towns and cities, a time when Americans relied heavily on commercial photographers to document many of life's most important moments, from birth to death and everything in between. It is not surprising, then, that our earliest photographic evidence of Providence Canyon dates from this period. But the Humber photo seems little more than a keepsake and curio, even though clearly it aimed to capture the impressive erosional formations that towered over the seated figures. There is little evidence before 1900, photographic or otherwise, that these gullies had yet captured the imaginations of an audience beyond the local.[26]

The local papers were remarkably silent about the gullies developing in the county during the late nineteenth century. Occasionally, though, the scale of the county's erosion problem peeked through, as it did in an 1887 item in the *Lumpkin Independent*, in a conversation that many growing towns and cities were then having about another pressing environmental problem: the need for municipal sanitation. "The needed improvements, every one knows, cannot all be done at once," the paper noted, "but cannot the city fathers extend their sphere of usefulness so far as to have removed a portion at least, of the immense quantity of rubbish, garbage, etc. which has for years been accumulating in the rear of business houses, and in some places, just off the side walks about the Public Square?" "In the vicinity of the heart of town," the paper continued, offering a clever solution for Lumpkin's mounting trash problem, "there are yawning, abysmal gullies, whose ever-widening and deepening mouths are ever approaching nearer and nearer to desirable localities,—

Figure 4. "Providence Caves" in 1893. R. T. Humber arranged for a commercial photographer to take this photo of what was then known as Providence Caves. For a sense of scale, notice the three figures in the lower right of the image. Courtesy of the Georgia Archives, Vanishing Georgia Collection, stw002.

steadily, but stealthily, and silently approaching—and into their insatiate throats might be dumped ton after ton of this accumulated rubbish." This suggestion, made with remarkable literary flair, aimed to solve two environmental problems at once: to rid the city of its growing piles of refuse while also slowing gully development and checking "their ostensible purpose of engulfing the town."[27] But not all of the locals, it seems, saw these gullies as yawning and abysmal. A few years later, just two days after Thanksgiving in 1889 and several years before the Humber photograph, the same paper noted that a "party of young men and young ladies returned their thanks at Providence, where they spent the day delightfully in exploring the deep caves and roaming through autumn forests."[28] As the rare stories such as these make clear, the gullies of Stewart County were enormous and increasingly problematic for some local residents by the late 1800s, while for others they were becoming places of local leisure. But they were not yet attracting much outside attention.

All of that changed in the early twentieth century.

CHAPTER TWO

The Most Picturesque Features of the Coastal Plain

Geologists Arrive at Providence Canyon

BY THE EARLY YEARS of the twentieth century, when its members first paid heed to Providence Canyon, the Geological Survey of Georgia was more than half a century old, and its mission and style had evolved considerably since its first incarnation. Between 1820 and the 1840s, during what a number of historians have called "the era of the state survey," many state governments created geological surveys and hired geologists and others trained in natural history to survey state lands for valuable resources.[1] In 1836, in line with that pattern, Georgia charged John Ruggles Cotting, an Ivy League–educated physician and Congregational minister, with making an initial geological survey of the state. The results of his survey, as of many geological surveys during this period, focused extensively on Georgia's soils, their agricultural capacities, and the various techniques for improving them.[2] Much of Georgia was still an agricultural frontier in the 1830s, and determining which lands would be best suited for agricultural development was an important state interest. Moreover, before the industrial revolution sparked enthusiasm for the large-scale mining of coal, iron, copper, and other useful ores, the greatest practical application of these surveys was to agriculture. As the historian Benjamin Cohen has shown, these antebellum surveys often built upon county-level agricultural reform efforts, were a critical aspect of the era's larger focus on internal improvements, and helped to bring a scientific approach to agriculture.[3] Although several of them made important contributions to the development of modern geology as a theoretical science, they were first and foremost applied exercises that worked at the intersection of science and agricultural improvement.

John Cotting visited Stewart County in the late 1830s, though he said nothing in his report about any incipient gullying. It was still a bit early for that. But in his larger report he did take the state's farmers to task for what he referred to as "the present ruinous and unscientific method of cultivating the soil." "*The worn out fields, the gullied hills and barren wastes* in every part of the State where the Georgia planter has placed his paralizing [sic] foot," Cotting complained, "are sad mementos of the effects of ignorance and prejudice on the part of our predecessors."[4] Cotting was in the southern surveying mainstream in emphasizing the prevalence of soil exhaustion, gullying, land abandonment, and the need for scientific expertise in rectifying the situation. South Carolina commissioned Edmund Ruffin, the famous Virginia reformer, to undertake its first survey in 1843, one that focused on making the state's agriculture more permanent and less destructive.[5] In Virginia, William Barton Rogers led a state geological survey between 1835 and 1842 that aimed to address the ruinous state of agriculture there.[6] And the state of Mississippi charged Eugene W. Hilgard, who would become one of the premier soil scientists in the United States in the late nineteenth and early twentieth centuries, with making a geological survey in the late 1850s. Hilgard's 1860 report, with its focus on soil classification and mapping, was a harbinger of soil surveys to come, and it too lamented the "melancholy waste of precious resources" by the state's farmers and planters. Throughout the South, then, state geological surveys were often concerned with correcting or improving agricultural practices that were, in Hilgard's words, "robbing the soil."[7]

While it might seem strange that geological surveys would pay so much attention to soils and agriculture, these early surveys occurred at a moment when many saw geology and agriculture as related fields. Indeed, many of these surveys could best be described as geological *and* agricultural. Most recognized that subsurface geology had an impact upon the productivity of soils, and many were intent on finding mineral deposits that would be useful soil amendments. Over the next half-century, however, soil as a natural body would come to assume an identity distinct (if not entirely separate) from subsurface geology, and it would get its own scientific discipline in the process. Providence Canyon's competing interwar interpretations would be the product of that expanding fault line.

The Civil War and its disruptive aftermath put a temporary halt to scientific surveying in Georgia and throughout the South, and when similar efforts developed again a decade or so after the war, they became increasingly institutionalized, professionalized, and focused on subsurface geology rather than on soils and their place in agriculture. To some degree this new focus on the subsurface was a reflection of a new interest in mineral riches, an interest fueled by the midcentury California Gold Rush and by the later mineral exploration of the American West to serve the second industrial revolution. Mining became both lucrative and essential to industrial processes, and geologists were happy to serve this new master. Where for much of the early nineteenth century the earth's greatest riches were believed to be in its shallow surface layer, the industrial era encouraged Americans to search more deeply.[8]

The split between soils and subsurface geology was also the result of the growth, at both the federal and state levels, of new departments of agriculture, which took over the tasks of haranguing farmers and dispensing advice about soils. A wartime Congress created the U.S. Department of Agriculture in 1862, and Georgia followed suit by creating its own department of agriculture in 1874. Importantly, the latter development was driven by the commercial fertilizer boom that fed the southern staple crop economy after the war. I will return to this story in a later chapter, but for now there are a couple of germane points. First, the widespread availability of commercial fertilizers after the Civil War drew attention away from the scientific study of in situ soils and the need to produce organic fertility within the bounds of the farm. Soil fertility could now be purchased and applied. Second, the unregulated market in commercial fertilizers resulted in the circulation of many products of questionable efficacy. As a result, when Georgia created its state department of agriculture in 1874, its function was almost solely to inspect and analyze commercial fertilizers so as to protect the state's farmers from unscrupulous salesmen and their often-fraudulent products. The same was true of many other fledgling state departments of agriculture, particularly in the South, where commercial fertilizer use was heavy. Studying soils themselves, at least beyond their discrete chemical constituents, had briefly become of secondary importance.[9]

Major changes occurring during the nineteenth century in the science of geology also pushed the Geological Survey of Georgia in new directions. Those changes were complex and followed on geological debates of the previous century, but we can still make several generalizations about them. First and most important, it was during the late eighteenth and early nineteenth centuries that concepts of geological deep time became widely discussed and debated. Indeed, the common understanding that the earth was in fact millions (and, later, billions) of years old, instead of thousands, was one of the signature intellectual achievements of nineteenth-century natural science. Although geologists still debated exactly what the major forces were that had shaped the earth's surface, the deep historicism of modern geology opened whole new fields of compelling scientific questions. For many, as we will see, it also created a new sense of sublimity. Moreover, developments in the realm of biology soon found their way into geological study, as geologists became interested in using fossils to date geological strata and to make sense of the history of life on Earth. As they built a layered picture of the earth's history, geologists traveled the globe to study stratigraphy and correlate their far-flung observations with those made closer to home. Coincidentally, another enormous gully in Georgia played a small role in these momentous changes in geological science. Its story is worth a quick detour.[10]

The best-known gully in the mid-nineteenth-century South was near Milledgeville, Georgia, on the south-central Piedmont. Known locally as Hall Gully, after T. M. Hall, the landowner on whose property it developed, it later came to be known as Lyell Gully, renamed for the eminent British geologist, Sir Charles Lyell, who visited the gully, broadcast information about its existence, and drew several lessons from it. Lyell was best known for his *Principles of Geology*, a three-volume text initially published between 1830 and 1833, which did for geology what Charles Darwin, himself heavily influenced by Lyell, would do for biology a quarter-century later. The eminent evolutionary biologist and historian of science Stephen J. Gould called *Principles* "the most famous geological book ever written."[11] These volumes appeared just as early settlers were moving into Stewart County. A proponent of the theory known as uniformitarianism, Lyell argued that the geological state of the planet was a product of processes, observable in the present, that had occurred across huge spans of time. As he put it in an early subtitle of *Principles*, his was "an attempt to explain the former changes of the Earth's surface by reference to causes now in operation." More than any other nineteenth-century geologist, it was Lyell who elucidated and popularized the idea of deep geological time. His uniformitarian ideas stood in contrast to reigning catastrophist theories of the era, which postulated that the earth's geology resulted from a series of punctuated and monumental upheavals, events that many catastrophists ascribed to the workings of God and that some still attempted to fit within a biblical timescale. Among these catastrophic events, many believed Noah's Flood was preeminent.[12] In the mid-nineteenth century these two camps were divided, and in the uniformitarian firmament Lyell was a star.

Lyell's encounter with the Milledgeville gully was an outgrowth of his travels in North America in the 1840s. Invited to give a series of lectures at the Lowell Institute in Massachusetts in 1841 and again in 1845, Lyell used these opportunities to explore eastern North America, recounting his experiences in two published travelogues.[13] Lyell was particularly keen to examine North American strata and their fossil record to see how well they corresponded with those of Europe, but he also searched for contemporary everyday processes, what he came to call "modern changes," that he could use to illustrate how geological formations had developed in the deep past across longer timescales. While his uniformitarian views emphasized slow and incremental change, Lyell recognized the power of storms, floods, earthquakes, and other forces to produce "low frequency, high magnitude" transformations.[14] To that end, in the literally titled *A Second Visit to the United States of America*, he chronicled several instances of severe gully erosion in the South, including the Milledgeville gully that came to bear his name. Lyell was not above using his travels and observations to criticize southern farming. He noted, for instance, that many of the Piedmont and Coastal Plain rivers that once ran crystal clear had, by the time of his visit, become turbid, the result of the land cover and land use changes wrought by settlers upstream. "I shall have occasion," he noted in volume one of his *Second Visit*, "to recur to this subject, when speaking of some recently-formed ravines of great depth and width in the red mud of the upland country near Milledgeville in Georgia."[15] There was certainly a sniff of disapproval there. But when he came upon the eponymous gully, Lyell generally resisted editorializing about southerners and their land use practices. Rather, he viewed the spectacle as evidence for his geological theories.

On his trips to the United States, Lyell attempted to meet as many prominent American scientific figures, and to visit as many important sites of geological research, as he could. He was impressed by the various state geological surveys, and he made a point of seeking out key figures in those efforts.[16] It was just such a mission that brought Lyell up to Milledgeville, then the state capital, in January 1846. Lyell was intent on meeting the Georgia state geologist, John Cotting, who had just published his report of the first Geological Survey of Georgia. During his visit, Lyell and Cotting went on an "excursion into the neighborhood of the capital" to make some field observations. One "most singular phenomenon" that Lyell noted in the Milledgeville area was "the depth to which the gneiss and mica schist have decomposed in situ." "Some very instructive sections of the disintegrated rocks," he continued, "have been laid open in the precipices of recently formed ravines." As his language indicated, these "ravines" mostly interested Lyell from a geological standpoint. He particularly commented on the state geological survey's identification of these as tertiary deposits when he thought they were older strata that had decomposed on site (Lyell was right). He then went on to describe the most extreme of these ravines, which lay at "a distance of three miles and a half due west of town, on the direct road to Macon." To visually portray the magnitude of this ravine of modern origin, Lyell included in his *Second Visit* an evocative woodcut of the spectacle. "Twenty years ago it had no existence," Lyell wrote of the deep "ravine," but then settlers arrived and cleared the land of its cover, and "in the course of twenty years, a chasm measuring no less than 55 feet in depth, 300 yards in length, and varying in width from 20 to 180 feet was the result." Later scholars, in carefully reconstructing the erosion history of Lyell Gully, have challenged the accuracy of Lyell's estimations. The gully when he saw it was likely only two-thirds as long as Lyell believed it to be. But it was nonetheless a powerful example of the very "modern changes" that were so crucial to supporting Lyell's uniformitarian views. As importantly, it gave him an opportunity to examine "the perpendicular walls of the great chasm" and the various layers and fossils contained therein. While he recognized that the gully was a "source of serious inconvenience and loss" for local residents, and that human land uses had contributed to its formation, he focused on the geological details and possibilities.[17]

The gully that came to be known as Lyell Gully was neither the only one nor even the most impressive one that Lyell witnessed on his travels in the South. As he headed west through Georgia and Alabama in early 1846, Lyell noted several other areas of more substantial gullying, including near Tuscaloosa, Alabama. In these areas, he found gully specimens that he ventured were more than seventy feet deep in places. But his most tantalizing observation, at least in the context of Providence Canyon's history, came during a stopover in Columbus, Georgia, an area that he toured in the company of several local naturalists. Lyell fossil-hunted along the Chattahoochee with Dr. Samuel Boykin, and a local named "Mr. Pond" took him up Upatoi Creek, a tributary of the Chattahoochee that today cuts across Fort Benning. In the wake of these excursions, he observed of Columbus and its surroundings: "Here, as at Milledgeville, the clearing away of the woods, where these Creek Indians once pursued their game, has caused the soil, previously level and unbroken, to be cut into by torrents, so that deep gulleys [sic] may every where be seen."[18] Lyell likely did not have any knowledge of the areas then beginning to erode in Stewart County just to the south, but he did provide evidence of early gullying that was well within the home range of Providence Canyon.

The spectacles of the Milledgeville gully and some of the other gullies of the South so impressed Lyell that he included a brief discussion of them in the ninth edition of *Principles*, which appeared in 1853. "When travelling in Georgia and Alabama, in 1846," he wrote, "I saw in both those States the commencement of hundreds of valleys in places where the native forest had recently been removed." Much of the rest of his discussion here was taken verbatim from *A Second Visit*, and he also included again the woodcut illustration of the Milledgeville gully. But his use of the naturalizing term "valleys" is an instructive one, as it better served his uniformitarian goals. Indeed, the section in which he discussed the Lyell Gully was titled "Newly Formed Valleys."[19] By discussing this "valley," his intent was not to use the spectacle of giant gullies to reprimand southern farmers for their erosive land use, as had Cotting in his survey report, or to make a moral distinction between geological and anthropogenic erosion, though he recognized the difference. Rather, he made

Figure 5. Woodcut of Lyell Gully. This woodcut by Charles Lyell originally appeared in his *Travels in North America* (1845).

gully erosion a central topic of modern geological study, divorcing it from the soil degradation narratives of antebellum reformers. In doing so, he helped to open up new scientific and aesthetic possibilities for appreciating landscapes of erosion that would be critical to later debates about Providence Canyon's meaning. As the art historian Barbara Novak has noted, by the 1840s "Lyell was well known and respected in American artistic and literary circles." Novak suggests that Lyell's uniformitarianism, by extending the age of the earth in the public's mind, informed the growing belief that landscapes revelatory of deep time were worthy of a new sort of aesthetic appreciation, what we might call the geological sublime.[20]

Such an appreciation, however, would take more than half a century to flower in Georgia, which brings us back to the Geological Survey of Georgia. After John Cotting's efforts in the 1830s and 1840s, the state of Georgia did not commission another geological survey until 1874, which was, tellingly, the same year it created its department of agriculture. This particular survey lasted seven years and was overseen by George Little. Apropos of his name, Little did not publish much, but in a small 1878 booklet on the specimens collected during the second geological survey he did make an oblique reference to substantial soil erosion in and around Stewart County. In a discussion of the state's deposits from the Cretaceous period, to which the eroded subsoils at Providence Canyon date, he noted: "This formation has but a small representation in Georgia. Its beds are exposed in a narrow belt extending from Columbus southward, nearly to Fort Gaines. . . . The country is unusually broken from denudation."[21] Fort Gaines is a town on the Chattahoochee River about forty miles south of Providence Canyon, near the southern boundary of the Fall Line Hills, so Little's description maps quite well onto the area of extensive gullying in the region including and surrounding Stewart County. But Little, like Lyell before him and true to the state survey's drift, seemed interested in this denuded country for its revelatory look at geological strata and the access it provided to Cretaceous fossils. Unlike Cotting, he was not interested in surveying soils or in moralizing about poor agricultural land use. The Geological Survey of Georgia had changed by the late 1870s.

As the second incarnation of the Geological Survey of Georgia came and went, the federal government formed a geological survey of its own. In 1879, Congress created the U.S. Geological Survey (USGS), building upon and consolidating the work of several western surveys. The main purpose of the new agency was to classify the lands of the vast public domain of the American West and to assess their geology, mineral resources, and, to a limited degree, their agricultural possibilities. The arid West, unlike the humid East with its cloaking vegetation, was a veritable amphitheater for viewing geological processes that had played out across deep time. There was also enthusiasm in the region for using scientific surveying to discover valuable mineral resources, as it was the treasures of the earth that had drawn so many Americans to the West in the mid- to late nineteenth century. But the USGS also employed scientists such as Grove Karl Gilbert and Clarence Dutton, who were true natural historians interested in explaining the mysteries of geological change over time in this bare region, and they helped to raise American earth sciences to a position of esteem in the world scientific community by the end of the century. John Wesley Powell, who took over the USGS in 1881, was an important figure in opening more space for basic research within the USGS. Ambitiously, he launched a massive mapping project to chart the topography of the entire country in a series of uniform and now-familiar USGS topographical maps. But in its attention to topography and geology, the USGS would be much less engaged with the soil than the earlier state geological surveys had been. "Despite the fact that soil was a prominent part of the earth's history and structure and of vital interest to humans," the environmental historian Donald Worster noted, "the [U.S.] Geological Survey paid it little attention."[22]

This was not entirely for a lack of trying. Unlike his immediate predecessor, Clarence King, and the USGS directors that followed him, all of whom made mining and mineral resources their central practical concern, Powell was keenly interested in agricultural reform. In the first years of the USGS's existence, Eugene Hilgard, who had completed his remarkable geological and agricultural survey of Mississippi in 1860 and was just then finishing a landmark report on cotton production in the United States, to be part of the 1880 federal census, urged the USGS to conduct a proper "agricultural survey" of the West, one that would include close attention to soils. Hilgard had left Mississippi in 1873 and, after a brief stint at the University of Michigan, had become a professor at the University of California at Berkeley in 1875, where he researched the soils of the arid regions. He thought that the USGS was the perfect vehicle for helping farmers, and he found a kindred spirit in Powell. In the mid-1880s, Powell tried to hire Hilgard to lead a proposed USGS division on "agricultural geology," but Hilgard demurred. One result was that soil science never gained the traction in the USGS that Hilgard hoped it would. This would not be the last time that Hilgard turned down a job he probably should have taken. In 1891, the USGS, as part of its annual report, did publish a treatise by eminent Harvard geologist and paleontologist Nathaniel Southgate Shaler, titled *The Origin and Nature of Soils*, which included a section on erosion as a particularly southern problem. But when Powell resigned under pressure in 1894, hope that the USGS would be a force for agricultural land reform left with him. The U.S. Department of Agriculture (USDA), as we will see, took a firm hold over soils research in the 1890s, and the USGS, in turn, focused on applied research in geology, its ambitious topographical mapping project, studies in irrigation, and the identification of valuable mineral resources. Outside of the public lands of the West, its main theater of operation, the USGS engaged with and helped to rejuvenate state geological surveys.[23] The State of Georgia, buoyed by the example of the USGS and aided by its cooperative spirit, created a permanent Geological Survey of Georgia in 1889, one focused on identifying valuable mineral resources in the state.[24]

It was in this context that, in the early twentieth century, Providence Canyon first appeared prominently in several Geological Survey of Georgia publications. An image of Providence Canyon (or perhaps of a neighboring gully) served as the frontispiece in Assistant State Geologist Otto Veatch's technical bulletin, *Second Report on the Clay Deposits of Georgia*, which the Geological Survey of Georgia published in 1909. The photograph apparently was included only for its stunning visual qualities, for nowhere in his lengthy study did Veatch discuss what he called, in his frontispiece caption, "Erosion Gullies in Upper Cretaceous Sands, Eight Miles West of Lumpkin, Stewart County, Georgia."[25] Save for those with an interest in the geographical distribution of economically useful clay deposits, the report was not particularly noteworthy. Two years later, though, in the *Preliminary Report on the Geology of the Coastal Plain of Georgia*, Veatch and his coauthor, Lloyd W. Stephenson of the U.S. Geological Survey (an example of the state-federal cooperation that the USGS fostered), discussed Stewart County's gullies at greater length. "The deep gullies, also known as 'washes' and 'caves,' which appear in this region," they wrote of the western portion of the state's Fall Line Hills, "are worthy of note." "Perhaps the largest and most picturesque of these gullies are located west and north of Lumpkin, Stewart County," they noted. "The gullies at Providence, eight miles west of Lumpkin, are 100 to 175 feet in depth, and from 200 yards to one-fourth mile in length."[26] While they noted that human land use had played a role in creating these new landforms, Veatch and Stephenson focused their comments, somewhat surprisingly, on the scenic qualities of these erosional formations. "These gullies," they concluded, "are the most picturesque features of the Coastal Plain, presenting curious erosion forms, pinnacles, 'islands,' or blocks of strata cut off from the adjacent land by erosion, and sharp, serrated ridges, while the bright red and white sands and the dark green pine tops present a vivid color contrast." They then instructed, rather matter-of-factly, that many similar "deep washes or gullies appear in Quitman, Webster, Marion, Crawford, Houston, Twiggs, Wilkinson, and Washington counties," counties that sweep east through the Fall Line Hills most of the way across the state. They made particular note of the "deep washes" at a place

Figure 6. Frontispiece from *Second Report on the Clay Deposits of Georgia*, 1909.

called Rich Hill, in Crawford County, which they thought was, "next to the Providence gullies, probably the most picturesque." They also mentioned Lyell Gully as a comparable phenomenon.[27]

One of the earliest explicit discussions of what would become known as Providence Canyon, this was also the first formulation that I have found of the place as aesthetically pleasing. It is no mere coincidence that such sentiment came from geologists. These representatives of the Geological Survey of Georgia and USGS were primarily interested in the gullied lands of Stewart and adjoining counties for what they might reveal about the geological history and mineral resources of the region. They were not directly concerned with the soil as a discrete natural resource or with the farming practices that seemed to be undermining its utility. In that sense, they were the heirs of Lyell, not Cotting. But they also began to mark these gullies as a scenic wonder, a gesture that other Georgians would soon embrace with greater enthusiasm and less modesty. In doing so, they were part of a larger aesthetic enterprise associated with geology at the beginning of the twentieth century. It was at this intersection between maturing geological science and a growing scenic appreciation of grand landscapes of erosion that Providence Canyon's moniker as Georgia's "Little Grand Canyon" became tenable, if wildly hyperbolic. It is not clear when Providence Canyon first acquired this flattering sobriquet, but before the twentieth century, such a claim of kinship, as a method for attracting tourists, would have made little sense. After World War I, though, it proved a powerful comparison.

The Grand Canyon of the Colorado itself underwent a remarkable shift in interpretation and appreciation in the late nineteenth century. When European and American explorers first encountered the Grand Canyon and the other canyon lands of the Colorado Plateau, they did not immediately recognize sublime masterpieces or natural wonders worthy of protection. What was remarkable about the reactions of early explorers, given our current esteem for the Grand Canyon as a place of self-evident beauty and grandeur, was their silence about the place. The Spanish explorers who encountered the canyons of the Colorado in the sixteenth and eighteenth centuries were, according to historian Stephen Pyne, "remarkable for their conservatism, their silence, their stubborn incuriosity about anything outside their prescribed agenda."[28] Even antebellum fur trappers saw these landscapes as obstacles or, at best, curiosities.

So how did the canyon become grand? How was it that landscapes of massive geological erosion came to be understood as scenic or even sublime? A confluence of developments in the late nineteenth century created the cultural canyon. The Grand Canyon became U.S. territory at midcentury, and it was soon the subject of federal exploration, most heroically when John Wesley Powell and his party boated the length of the Colorado River through canyon country in 1869 (and then, a second time, in the early 1870s). As a place of heroic exploration, the Grand Canyon came to have national cultural meaning. Moreover, in the wake of a divisive Civil War that had nearly destroyed the nation, many turned to the dramatic landscapes of the West for their unifying and redemptive qualities. Powell, the veteran who had lost an arm at Shiloh, was the very embodiment of this impulse. As importantly, American chauvinists began to favorably compare the grand nature of the American West with the cultured landscapes of Europe, seeing in the wild sublimity of places such as the Grand Canyon an important forge of national identity. American landscape painters and photographers also began producing images of the Grand Canyon and other western spectacles, many of which would soon be enshrined in national parks, that gave aesthetic consumers new perspectives on these places. Frontier romanticism and the elevation of a domesticated, Christianized sense of the sublime suddenly made the threatening and confusing spectacles of western geological erosion into sites of awe and pilgrimage. Finally, and most importantly, the historicism of modern geology allowed scientists to transform the canyon into a wondrous cross-section of the earth's deep natural history. Powell himself had gotten this started with his 1875 report, *Exploration of the Colorado River of the West*, whose attention to the geological processes of erosion and uplift that had created the canyon gave his adventure the aura of a trip back through deep time. Together, then, nationalism, heroic exploration, the arts, and modern geological science signified the Grand Canyon as a site of tremendous meaning and importance, of deep antiquity even.[29]

By the first decade of the twentieth century, the Grand Canyon had attained monumental status. So impressed was President Theodore Roosevelt when he visited in 1903 that he later protected the Grand Canyon as a national monument, using the new executive powers accorded him by the Antiquities Act of 1906, which allowed the president to protect as national monuments any areas within the public domain that were of scientific, archaeological, or cultural importance. Much of the area was already a forest reserve, but national monument status offered greater protection. American tourists, delivered first by the railroad, which arrived in 1901, and then by their automobiles, began flocking to see the spectacle and appreciate its aesthetic grandeur. In 1919, the Grand Canyon became a national park, its place in the pantheon of American natural wonders secured. As the interwar era began, then, the Grand Canyon was widely available as a scenic point of comparison.[30]

The favorable aesthetic judgment that Otto Veatch and Lloyd Stephenson offered of the "gullies at Providence" in 1911, though modest, would have been more difficult to arrive at without the historical processes that remade the Grand Canyon and other western landscapes of erosion into scenic wonders. As geological surveys became more singularly geological, and as the science of geology abandoned the soil for deeper strata and their temporal revelations, geologists helped to provide the intellectual and cultural tools to remake landscapes of erosion into compelling natural scenery. It is not surprising, then, that geologists in the early 1910s found the gullies of Stewart County "picturesque." But, of course, Providence Canyon was not the Grand Canyon. It was and is *far* smaller, it is of *much* more recent origin, and, perhaps most perplexingly, it is a human artifact. These differences mattered. As a result, while tourism to the Grand Canyon and other western landscapes of erosion boomed during the early decades of the twentieth century, a period when boosters urged Americans to eschew European travel and "See America First," Providence Canyon languished in a remote corner of the South.[31] It would take an additional set of historical developments before Providence Canyon's admirers could hope to convince a larger public that Georgia's "Little Grand Canyon" was worth the trip.

CHAPTER THREE

Rough, Gullied Land
Soil Scientists Arrive at Providence Canyon

IN THE IMMEDIATE WAKE of the scenic compliment that the geologists Veatch and Stephenson paid to "the gullies at Providence," another group of scientific surveyors arrived in Stewart County and began to frame a different perspective on its spectacular erosion. In 1912, the National Cooperative Soil Survey charged David Long, of the Georgia State College of Agriculture, and M. W. Beck, a USDA soil surveyor, with making a survey of Stewart County's soils. A year later, two more USDA officials, E. C. Hall and W. W. Burdette, joined the effort.[1] The National Cooperative Soil Survey (hereafter, the Soil Survey) was part of an ambitious federal-state effort, begun in 1899, to map the nation's soil resources, mostly at the county level. The scientists who worked for the Soil Survey were not initially charged with studying the causes and extent of anthropogenic soil erosion, but the erosion they witnessed during the first two decades of the twentieth century, particularly in portions of the South, slowly built a case for a modern program of soil conservation that would come to fruition in the late 1920s and 1930s, well before the Dust Bowl dramatized the need for such a program.[2] The soil survey of Stewart County played a pivotal role in that development, and it laid the groundwork for a conceptualization of Providence Canyon as the apotheosis of southern soil abuse.

This chapter examines the forces that led to the Soil Survey's creation in 1899 and its arrival in Stewart County, Georgia, in the years before World War I. It also focuses on how the Soil Survey, and the U.S. Bureau of Soils more generally, contributed in curious and indirect ways to the rise of a soil conservation movement in the United States that would make Providence Canyon famous in the 1930s. Two individuals are central to this story. The first and better known is Hugh Hammond Bennett, the founding father of American soil conservation, who headed the Soil Conservation Service from its creation in 1935 until his retirement in 1951. Bennett has long been among the pantheon of American conservationists, but scholars have been slow to appreciate how his early work as a soil surveyor contributed to his later advocacy for soil conservation. The second central character is Milton Whitney, the long-time chief of the U.S. Bureau of Soils. It was Whitney who created the Soil Survey, which would eventually cobble together a vivid picture of the nation's soil erosion problems. But, surprisingly, Whitney was indifferent to the problems of soil exhaustion and soil erosion, and his indifference, as the head of the most important federal agency for soils research, had a profound impact on debates over soil conservation and permanent agriculture in the early twentieth century. We cannot understand the development of the modern erosion consciousness that Bennett embodied, and the central place that Providence Canyon played in that emergence, without attending to the unlikely influence of Milton Whitney.[3]

The National Cooperative Soil Survey had roots in a nineteenth-century debate over what constituted soil fertility, and it would come into being as a result of Milton Whitney's own peculiar contribution to that debate. At the beginning of the century, the "humus theory" of soil fertility dominated scientific discussion. Humus theorists argued that fertile soils were rich in vital organic matter, or "humus," from which plants got their nutrition. The key to maintaining soil fertility in the face of crop production, they insisted, was the addition of organic matter through various forms of manuring—not only the use of animal dung but also green manures such as leguminous cover crops plowed into the soil. Sir Humphrey Davy's landmark 1813 book, *Elements of Agricultural Chemistry*, popularized the humus theory and proved influential in American agricultural circles. In the decades that followed its appearance, agricultural reformers, bemoaning the "worn-out" soils that American farmers had created all along the eastern seaboard, experimented with and advocated for various organic manures and the husbandry methods necessary for raising them on the farm.[4]

By 1840, however, several inconsistencies in the humus theory came to plague scientists, farmers, and reformers, among them how minerals affected plant growth, how inorganic soil additives such as marl and gypsum worked, and how plants got the nitrogen they needed. The German chemist Justus von Liebig stepped into the breach to offer a critique of the humus theory and a new approach to soil fertility that emphasized mineral inputs and chemical analysis. The timing of his 1840 treatise, *Organic Chemistry and Its Applications to Agriculture and Physiology*, was impeccable. Not only were American farmers and scientists open to chemical explanations of soil fertility, but Liebig's emphasis on the necessity of just a handful of mineral elements for maintaining soils meshed beautifully with the growing availability of commercial fertilizers over the next several decades. The implications of Liebig's approach were both wonderfully and dangerously reductive. While he did not deny the usefulness of organic manures in rebuilding soil fertility, he believed that their efficacy lay in a few essential chemical elements, not in the restoration of humus as a vital organic body. Liebig suggested that farmers simply needed to chemically test their soils to see what elements were missing and then add them accordingly. Indeed, Liebig imagined a future in which fertilizers would be denatured and purely chemical, produced in laboratories and manufacturing plants. Liebig also argued, incorrectly, that plants got the nitrogen that they needed from atmospheric deposition, and that most soils thus had sufficient nitrogen. This was perhaps the most important mistake he made, one that diverted farmers' attention from organic manures and leguminous crops. It would be another half-century before scientists demonstrated that soil microbes were the key actors in fixing atmospheric nitrogen in the soil, a discovery that would renovate the reputation of humus as microbial habitat. Liebig's reconceptualization of soil-plant relationships, which dominated the debate at midcentury, represented a shift from organic to mechanistic understandings of soil fertility and laid the conceptual foundation for the modern chemical fertilizer industry.[5]

As a result of Liebig's work, and a wave of American enthusiasm for it, questions about soil fertility became the province of chemistry, at least for a time. As the soil scientist David Montgomery has put it, "Faith in the power of chemicals to catalyze plant growth replaced agricultural husbandry and made both crop rotations and the idea of adapting agricultural methods to the land seem quaint."[6] During the 1840s and early 1850s, a "sudden craze over soil analysis," to use historian of science Margaret Rossiter's phrase, overshadowed the work of the state geological surveys, which waned in influence during these years—although some of the surveys incorporated chemical soil analysis as part of their charge.[7] A focus on soil chemistry and a rejection of the humus theory thus moved the search for the secrets of soil fertility out of the barnyard and into the lab.

During the second half of the nineteenth century, however, there emerged a growing sense among both farmers and scientists that Liebig's chemical theory had important flaws. Those shortcomings were many and complex, but the simplest was that Liebig's central promise—that soils could be chemically analyzed to assess their deficiencies and then amended with precise amounts of certain mineral elements to create an ideal balance of chemical fertility and thus uniform plant productivity—often failed in practice. Some of the problems came in the imprecision of chemical analyses, the varying quality of fertilizer amendments, and even the vagaries of human application. But farmers also discovered that soils that were identical in their chemical profiles, but that varied in other ways, responded to fertilizer inputs differently. And so agricultural scientists set out to discover why. As they did so, and as geologists ceded them this scientific territory, they began to professionalize the modern field of soil science and to make a central place for soils research in growing federal and state agricultural bureaucracies.[8]

Liebig's bold revisionist theory, however incorrect in some of its details, created questions and problems that lent themselves well to scientific research. In fact, the longest running set of experiments on soil fertility and its maintenance, the "Classical Experiments" at Rothamsted in England, began in 1843 in response to some of the perceived shortcomings of Liebig's theories.[9] During the second half of the century, no single unified theory of soil fertility emerged to replace Liebig's. Instead, soil scientists revisited organic theories of fertility, reexamined how inorganic amendments functioned to increase soil productivity, turned their attention to newer questions about the role of climate in shaping soils, delved into the importance of soil biology and soil physics to plant productivity, and slowly unlocked the mysteries of nitrogen fixation. As Rossiter has argued of the soil science of the late nineteenth century, "Liebig's simple and strictly chemical view had been rejected for a more complex and comprehensive one of plant-soil relationships."[10] During this same period, scientists at American universities and newly created state research facilities took up these questions, all the while working to convince the farming public that their results could be useful. In 1887, Congress passed the Hatch Act, which created a series of state-level agricultural experiment stations. In some cases, newly available federal funding and support went to state experiment stations already up and running, but the Hatch Act helped to spread the experiment station model throughout the nation and built a case for stronger federal coordination of agricultural research. Soils research was a major beneficiary.[11]

Then, in the 1890s, a new reductionism emerged, intersecting in powerful ways with the federal institutionalization of soil science. The key figure in these developments was a young and ambitious soil investigator, Milton Whitney, who damned complexity in favor of one big idea. Whitney, born in Maryland in 1860, had studied chemistry at Johns Hopkins University without earning a degree. Then, in 1883, he took a position with the Connecticut Agricultural Experiment Station, the nation's first, where he worked as an assistant to the eminent soil chemist Samuel Johnson. After this critical internship, Whitney went on to positions at agricultural experiment stations in North Carolina, South Carolina, and Maryland. He was one among a growing group of soil scientists for whom experiment station work was formative. Fortuitously, Whitney's former boss in North Carolina, Charles Dabney, became an assistant secretary of agriculture in charge of the department's scientific work in

1894. This happened just as John Wesley Powell was resigning from the USGS and soil science was fading from its purview. Moreover, Dabney accepted the position only after E. W. Hilgard had turned it down when it had been offered to him. Dabney then hired Whitney to head a new Division of Agricultural Soils within the USDA, a division that marked a commitment by the USDA to engage in soils research. Both Whitney and Dabney were keen on understanding the shortcomings of Liebig's chemical theory of soil fertility, and Whitney homed in on the role of soil texture. Noting that soils with the same chemical profiles but with different textures produced crops at different rates, he argued that texture was the key to soil productivity. Whitney was not alone in arguing for greater attention to physical factors in explanations of soil fertility, but he was unique in his single-mindedness about their importance. He boldly and incorrectly insisted that almost all soils had adequate nutrient stores to grow crops indefinitely, and that texture and the related factors of moisture and temperature determined the productivity of a soil. According to Whitney, there was no such thing as soil exhaustion, and there was no real need, from a nutritional standpoint at least, to fertilize soils.[12]

Whitney developed this remarkable position while he occupied a key position in the USDA's growing research bureaucracy. He headed the Division of Agricultural Soils, which was reorganized as the Division of Soils in 1897 and then became the Bureau of Soils in 1901, until his retirement in 1927, and he used that bully pulpit, such as it was, to diffuse his faulty soil texture theory at the expense of other theories of soil fertility.[13] One of the products of Whitney's insistence on texture's centrality was the creation of the Soil Survey, over which Whitney presided either directly or indirectly for the rest of his career. We will get to the details of that story soon. But another important product of Whitney's single-minded focus on soil texture, and his insistence that soil exhaustion was a myth, was the enmity of several other prominent soil scientists. Whitney and E. W. Hilgard feuded almost from their first contact in the 1890s, with Hilgard even going so far as to try to get Whitney removed from his USDA position.[14] But it was Whitney's run-ins with a couple of other soil scientists that had a more lasting impact on the shape of twentieth-century agricultural reform.

Those who strayed from the increasingly rigid Whitney orthodoxy found their career prospects limited within the Bureau of Soils. The best example here was a clash between Whitney and a soil scientist named Franklin Hiram King. King began his career as a professor of "agricultural physics" at the University of Wisconsin, the first such position in the country. Presumably impressed by his commitment to soil physics, Whitney hired King in 1902 to serve as chief of the Division of Soil Management in the Bureau of Soils.

Figure 7. Milton Whitney, chief of the Bureau of Soils, with a National Cooperative Soil Survey map. Travis P. Hignett Collection of Fixed Nitrogen Research Laboratory Photographs, Chemical Heritage Foundation Archives. Courtesy of the Chemical Heritage Foundation.

During his brief stint with the bureau, King got into trouble for daring to disagree with the soil fertility theories promulgated by Whitney. King believed that fertilizing soils was essential to maintaining key plant nutrients, and he conducted a series of studies for the Bureau of Soils that, he believed, demonstrated not only the efficacy of fertilization but also that Whitney was wrong to believe that soils had sufficient nutrients to grow crops indefinitely.[15] As a result of this apostasy, Whitney forced King's resignation in 1904 and refused to publish several of King's research bulletins.[16]

King's coerced resignation and the withholding of his research findings created a storm of protest within the soil science community. Hilgard led the charge with a scathing editorial in *Science* in which he railed against the "ill-founded hypotheses of the head of one of the most important bureaus in the Department of Agriculture." "It is impossible to conceive that in the twentieth century, and especially in a country claiming to be progressive *par excellence*,"

Hilgard fumed, "such a regime should be allowed to continue for any length of time."[17] Cyril Hopkins, a soil scientist at the University of Illinois, took up King's cause as well, not only in speeches and writings, but also in a formal letter of complaint sent to the secretary of agriculture.[18] "The erroneous teaching so widely promulgated by the federal Bureau of Soils," Hopkins later insisted, "is undoubtedly a most potent influence against the adoption of positive soil improvement in the United States, because it is disseminated from the position of highest authority." "Other peoples have ruined other lands," Hopkins concluded, "but in no other country has the powerful factor of government influence ever been used to encourage the farmers to ruin their lands."[19] Ultimately, despite these howls of protest, Whitney maintained the support of Secretary of Agriculture James Wilson and thus his control over soils research within the USDA.

For their part, King and Hopkins used their feud with Whitney to build a case for what they together called "permanent agriculture," a phrase that would become the centerpiece of agricultural reform efforts in the first half of the twentieth century. In 1909, King embarked upon a nine-month journey to Asia to study the longevity of agriculture there and the role that organic fertilizers played in that longevity. Upon his return to the United States, he wrote about his experiences in *Farmers of Forty Centuries; or, Permanent Agriculture in China, Korea, and Japan*, which appeared posthumously in 1911, with a preface by Liberty Hyde Bailey, the chief ideologue of the Country Life Movement.[20] Asian farmers, King observed, had maintained the fertility of their soils over several millennia while Americans had depleted the soil of vast swaths of the continent with remarkable rapidity. "We had gone from practices by which three generations had exhausted strong virgin fields," he observed of his transit from North America to Asia, "and we were coming to others still fertile after thirty centuries of cropping."[21] Asian farmers had achieved this feat, while supporting dense populations and without the assistance of commercial fertilizers, by being remarkably efficient stewards of their available sources of organic fertility. King observed Asian farmers applying amendments such as green and brown manures, canal mud, seaweed, and ashes to their fields to keep them fertile, and he marveled at their composting techniques. He also documented Asian practices of collecting and applying night soil (human waste) to fields to maintain fertility, and he lauded Asian farmers for the various sanitary methods that they employed to keep this a safe practice.[22] In comparison, King rued the American "waste of plant food materials through our modern systems of sewage disposal and other faulty practices."[23] King's portrait of Asian agricultural was highly romanticized. He paid little attention, for instance, to the substantial history of soil erosion in China, a topic that soil conservationist Walter Lowdermilk would soon explore in detail.[24] But his respectful study of labor-intensive Asian farming techniques was a paean to the virtues of organic fertilizers and a revival of a reverence for humus. Although he wrote nothing about Whitney's dismissal of such techniques, *Farmers of Forty Centuries* was a clear critique of the doctrine that had driven him from the Bureau of Soils.

Where King's celebration of "permanent agriculture" in Asia came as an implicit attack on Whitney's theories, Cyril Hopkins was unabashed in his public excoriation of the Whitney orthodoxy. Hopkins produced three books between 1910 and 1913, all of which openly rejected Whitney's theories while offering his own prescription for what he too called "permanent agriculture." The first volume in his trilogy of repudiation, *Soil Fertility and Permanent Agriculture* (1910), was a substantive textbook, while the third volume, *The Farm That Won't Wear Out* (1913), was a brief primer on the subject. In both, Hopkins, who had been influenced by the long-term field experiments at the Rothamsted, insisted that plants were demanding of soil nutrients and that certain elements needed to be continually added by farmers to avoid declining fertility. He was right, of course. More importantly, he consistently and repeatedly rejected the "radical theory" of Whitney "to the effect that soils do not wear out or become depleted by cultivation and cropping," and he worried that Whitney's theories were only extending a long history of soil-depleting agricultural practices. "The possible enormous and irreparable damage of such teaching lies in the fact that even our remaining supply of good land will ultimately be depleted by the present practices beyond the point of self-redemption, thus repeating the history of our abundant Eastern lands." In both books, Hopkins explicitly and repeatedly rejected the "pernicious, disproved and condemnable doctrines poured forth and spread abroad" by the Bureau of Soils.[25]

But it was Hopkins's second book, *The Story of the Soil* (1911), that was his most unusual. In a quixotic attempt to make his side of the soil fertility debate accessible to the general public, Hopkins wrote a novel about a young soil scientist named Percy Johnston who leaves the Midwest for the South, with the goal of buying a farm there and restoring its exhausted soils. The novel opens with a shocking scene. As Percy is chauffeured away from a southern train station by Adelaide, the daughter of the local farmer with whom he will be staying, two African American men jump out of the woods, pull Adelaide from the wagon, and attempt to rape her. In a Herculean show of strength, Percy single-handedly subdues the two attackers, marches them into town, and hands them over to the police, preserving Adelaide's honor. The next day, when Percy returns to provide his testimony, he finds the two men hanging from a tree in the courtyard square, the victims of a lynching. It is not entirely clear why Hopkins felt the need to begin with such an episode, though such scenes were a convention of Jim Crow–era literature. Hopkins was likely mimicking the work of writers such as Thomas Dixon, whose 1905 novel *The Clansman* inspired D. W. Griffith's *The Birth of a Nation* a decade later. After this initial scene of racial violence, however, most of the rest of the plot plays out as a series of pedantic conversations in which Percy explains to anyone who will listen how soil fertility works and how the U.S. Bureau of Soils had strayed dangerously from scientific truth.[26] The book even includes a fictional trip by Percy to Washington, D.C., during which he visits the Bureau of Soils and has an animated conversation with its unnamed director, a conversation that basically lays out the entire Whitney-Hopkins debate. Finally, while Percy does win over Adelaide in the end, there is an interlude during which she becomes engaged to a Professor Barstow from North Carolina who, to add insult to injury, is an adherent of Milton Whitney's theories. *The Story of the Soil*, though a failure as fiction, is the most thorough and unusual critique we have of Whitney's misguided views.

Together, King and Hopkins, in their hostility toward Whitney, gave birth to an ideal of "permanent agriculture" that would later assume a central place in the New Deal critique of wasteful American, and particularly southern, farming practices. It was a critique that Providence Canyon came to visually represent. I will return to the details of that critique later. For now, the important point is that this ideal of "permanent agriculture" arose early in the twentieth century not only as a response to the troubling history of American agricultural impermanence but also because the theories of the nation's leading government soil scientist were antithetical to a reckoning with those destructive patterns. As John Paull, a student of this controversy, has concluded: "The concept of 'permanent agriculture' was, at least partly, the outcome of an internecine war, between the Bureau of Soils of the United States Department of Agriculture (USDA) and two U.S. professors of agriculture."[27]

The work of Hopkins and King to promote permanent agriculture quietly undergirded the organic agriculture movement that emerged after World War II. In particular, *Farmers of Forty Centuries* enjoyed belated acclaim. Though initially ignored, it was reprinted in 1927 in England, where it sold well and influenced Sir Albert Howard, a founding figure of the organic agriculture movement.[28] Then, in 1949, Jerome Rodale, the writer and publisher who popularized organic farming in the United States, published a new U.S. edition. Thereafter, *Farmers of Forty Centuries* became a recognized classic in organic agriculture circles. Tellingly, the phrase "permanent agriculture" was replaced in King's subtitle with "organic farming" in later editions.[29] After

World War II, when synthetic fertilizers and chemical pesticides allowed farmers to achieve permanence at great environmental cost, the ideal of an "organic" agriculture made more sense than did "permanent agriculture," for it stood in stronger contrast to the world of chemical artifice that had redefined farming. The battle that King and Hopkins waged against Whitney thus had a lasting though little-appreciated influence on environmental critiques of modern agriculture. As it turns out, this flurry of early writings on "permanent agriculture" was not the only meaningful product of Whitney's misguided theories and closed-minded leadership.

An equally important environmental legacy of Whitney's leadership of the U.S. Bureau of Soils was the emergence of a soil conservation movement focused on erosion prevention. It too was shaped by Whitney's insistence that soil texture mattered above all else. Whitney believed that one of the keys to getting useful information to farmers about soil texture and its relation to agricultural productivity was to survey and map agricultural soils. As Soil Survey historian David Gardner has noted, "Whitney's preoccupation with the importance of soil texture was implicit in his belief that soil surveys, based upon the recognition of physical soil conditions, would provide the long-sought touchstone for the proper location of crop production."[30] And so, in 1899, the USDA and the states created the National Cooperative Soil Survey, whose early focus was on mapping soil texture with an eye toward proper land use. Whitney was largely unconcerned with soil erosion as an agricultural and environmental problem for the duration of his career. Nonetheless, his insistence on mapping soils by texture led others to the realization that soil erosion was a major agricultural problem in the United States.

The Soil Survey initially confronted and conceptualized soil erosion as a problem of scientific classification, not as a problem of land use. To understand that curious fact, we need to appreciate how the early soil survey worked. Before soil surveyors could coherently map the nation's soils, they needed to figure out how to classify them. There were some models for doing so. For most of the nineteenth century, even as geology and soil science had drifted apart professionally, soil classification remained fundamentally geological. To put it simply, surveyors had classified soils based upon the underlying parent rock from which they originated. But in practice, among those that worked the soil, classification usually relied upon imprecise generic terms, many of them based on texture, dominant vegetation, or how soils responded to cultivation. Among the southern vernacular categories there was "sandy land," "clay land," "limestone land," "freestone soil," "mulatto land," "black jack land," "crawfish land," and "buckshot land," to name a few.[31] With the rise of soil chemistry in the decades before the Civil War, soil classification came to be more about determining the types and amounts of nutrients to be found in particular soils. Whitney's focus on texture not only rejected chemical and other tests of soil fertility, but it also began the slow process of distancing soil classification from underlying geology. While Whitney continued to believe that underlying geology helped to determine the nature of soils, he was most interested in having surveyors identify and describe the texture of what soil scientists today call the surface horizon. And so the Soil Survey created its own classificatory system, one that grew by fits and starts during the Survey's first years.[32]

Was a soil sandy, silty, clayey, loamy, or something in between? Was it fine or coarse? These were the critical questions of the earliest soil surveys, and out of this emphasis on surface texture emerged the fundamental taxonomic unit of the early Soil Survey: the *soil type*. The soil type consisted of a binomial designation that usually combined location and texture. For example, one of the four initial surveys completed in 1899 was of Cecil County, Maryland, which straddles the top of the Chesapeake Bay. In that survey, investigators identified a group of soil types that they categorized by combining local place names with texture descriptions. Two of the soil types they identified were "Cecil loam" and "Cecil clay," with "Cecil" functioning as a geographical marker for Cecil County and "loam" and "clay" describing texture. The initial assumption was that Cecil soils were exclusive to Cecil County, and that similar soils in adjoining counties would have different names and thus be identified as different soil types. "The soil type was considered a local kind of soil during the first few years of the American soil survey program," soil scientist Roy Simonson has noted. "For the purposes of correlation, each survey was a universe unto itself."[33] This would not be a tenable system for long.

As the number of soil surveys multiplied, so did the number of soil types. There were more than four hundred soil types by 1902, too many for surveyors to remember, and that number threatened to grow exponentially as soil surveyors fanned out across the nation. The whole system risked collapsing under its own weight. If they were going to accompany soil type mapping with practical advice for farmers, surveyors recognized, soil types would have to be limited in number and geographically expansive. As a result, the place names attached to soil type binomials shifted from being markers of specific locations to being indicators of similar soils existing across larger regions, with the place name indicating where surveyors first identified the type. Out of this redefinition came a new and higher soil classification introduced in 1903: the *soil series*. Thus, as a soil series, Cecil soils could exist in places other than Cecil County, Maryland. Indeed, soil surveyors quickly recognized Cecil soils as one of the dominant soil series of the southern Piedmont, a physiographic region that soil surveyors soon determined to be one of thirteen North American *soil provinces*, yet another layer added to this evolving classificatory scheme. Soil series and provinces brought hierarchical order and standardization to early USDA soil type classification, allowing surveyors and farmers to correlate soil types across larger areas. In the coming decades, this entire classification system would undergo a major overhaul at the hands of Curtis F. Marbut, a soil scientist who took over the day-to-day operations of the Soil Survey in 1911. By the late 1920s, Marbut would come under the sway of the work of several Russian soil scientists who insisted that climate, topography, and biology shaped soils just as much as underlying geology. Indeed, it was their work that finally divorced soil science from geology. But during the early years of the Soil Survey, what mattered most to Milton Whitney at a practical level was how particular crops adapted to and grew in particular soil types.[34]

Along with this type-series-province taxonomy, another important classificatory concept crept into soil survey reports and maps during the Survey's first decade of operation: the *soil phase*. Surveyors developed the soil phase as a classificatory improvisation, a response to field observations of soil conditions that did not match defined soil types. Crucially, surveyors began to categorize soils affected by human-induced erosion by using "phase" distinctions. While several early soil surveys struggled with how to classify eroded soils, the first to make mapping distinctions based upon erosion in the same soil type was the "Yazoo Areas, Mississippi" report of 1901, which concluded that much of the area was "badly eroded," and that erosion made it difficult to correctly classify some of the region's soils. That same year, in a poignant moment of realization, soil surveyors working in Cary County, North Carolina, came to recognize that what they had called Cecil clay was actually an eroded "phase" of Cecil loam. The clay they identified was, in fact, a clay subsoil from which the loamy surface layer had washed. This was likely true for many of the other places where Cecil clay had been identified as a soil type. Within a decade, it became common for surveyors to identify and map "eroded phases" of particular soil types in their reports. Erosion thus captured the attention of soil surveyors as they struggled with a new system of soil classification. By identifying and mapping these eroded phases, soil surveyors inadvertently built a picture of how humans had altered the nation's soils by exposing them to the forces of erosion.[35]

That American agricultural soils had a long history of exhaustion and erosion, and that southern soils had suffered in particular, was not a new discovery of the Soil Survey. Concerns about soil exhaustion and erosion were already more than a century old by the early twentieth century, and they had waxed and waned over the course of the nineteenth century. Even Milton Whitney had had to contend with the reality of erosion early in his USDA employment, when he helped Charles Dabney prepare a Farmers' Bulletin titled *Washed Soils* in 1894.[36] That said, soil surveyors entered the field at the beginning of the twentieth century assuming that the soils they encountered and mapped were products of nature, not artifacts of human land use. Not only did the Soil Survey bring system, standardization, and a lot more data to soil erosion observations, then, but soil surveyors came to understand that the soils they were mapping had a history. Soil surveyors also began to suspect that what many had historically referred to as soil "exhaustion" or "worn-out lands," soils made less fertile by the nutrient demands of successive plantings of staple crops, may also have been the results of cumulative sheet erosion washing fertile soils downslope. The first steps in this direction were tentative, but the evidence quickly mounted. As federal and state soil surveyors covered more ground, noting how human activities had changed the nation's soils, erosion, which they had initially encountered as a problem of classification, grew to be a conservation concern.

For a young soil surveyor named Hugh Hammond Bennett, mapping the nation's soils was a radicalizing process. Bennett joined the Soil Survey in 1903, having just graduated from the University of North Carolina at Chapel Hill. Bennett's awareness of erosion as an agricultural problem predated his college education. He had grown up on a cotton farm in Anson County, North Carolina, where erosion was an obvious reality. Later in his life, Bennett recalled helping his father build a terrace system, a common technical response among conscientious farmers to the threat of erosion, on the family's farm in 1891, when he was only ten years old. But for Bennett, soil erosion seemed only a "local farm difficulty" until he joined the Soil Survey.[37]

During Bennett's early years with the Soil Survey, he traveled widely, often spending weeks in the field, lodging with local farmers or camping when the day was over. His first survey assignment brought him to Davidson County, Tennessee, in the spring and summer of 1903. Then the winter of 1903 found him in Appomattox County, Virginia. In early 1904, he surveyed in Macon County, Alabama, the home of Booker T. Washington's Tuskegee Institute, where he apparently met Washington—and where he noted that several of the county's soils were particularly prone to erosion. There is no indication whether Bennett met George Washington Carver, who had arrived in Macon County less than a decade earlier and had become quite familiar with its eroded and fatigued fields. Carver was then running the agricultural programs at Tuskegee and developing an interest in ways to aid the county's poor tenant farmers, most of whom were African American. Improving the soil with readily available organic amendments was a big part of Carver's vision. After completing the Macon survey, Bennett went to New York's Finger Lakes region for

Figure 8. Hugh Hammond Bennett. This photograph was taken when Bennett had become "chief" of the Soil Conservation Service and the nation's leading soil conservationist. Courtesy of the Natural Resources Conservation Service, USDA.

the summer before cooler weather chased him back down south to Selma, Alabama. In all of these places, Bennett witnessed farmers struggling with eroded and exhausted lands.[38]

It was a 1905 survey of Louisa County, Virginia, that opened Bennett's eyes to the deeper historical problem of erosion in the South. Louisa, a Piedmont county that had been farmed for almost two centuries, had garnered a reputation as a place with poor soils, and so Whitney asked Bennett to look into the problem. As Bennett later recalled, the crux of the problem became plain when he compared the soils beneath the county's few woodland areas that settlers had never cleared with the soils of long cultivated fields. Even on similar slopes, the woodland soils were rich, moist, and full of organic material, while the cultivated land had poor clay soils that baked to a hardpan in the summer sun. While most soil surveyors would have identified and mapped these as different soil types based upon their textural differences, Bennett suspected that the county's poor agricultural soils had once been as rich as its forest soils. As he later recalled it, this observation triggered an important realization: sheet erosion had been quietly carrying away fertile topsoil and robbing farmers of their most important resource. Much of the early twentieth-century Piedmont, with its exposed red clay soils, was an artifact of human-induced erosion. Later in life, when he had become the nation's preeminent soil conservationist, Bennett referred to this Louisa County survey as his road to Damascus. Whether it truly was such a point of change for him is hard to say, as the published survey report itself did not betray this revelation. Bennett maintained that his erosion findings and observations, initially included in the Louisa County report, "failed to survive editing in Washington," although when I examined the edited version of the report and the correspondence that went with it I found little evidence of Bennett's concerns being silenced. Whatever the case, the Louisa County survey was one of several that led Bennett to realize the importance of soil conservation in a region transformed by erosion.[39]

As the Soil Survey developed during the first decade of the twentieth century, and as soil erosion came to be a growing problem for surveyors in the field, it intersected with the Progressive conservation movement in important if limited ways. In the late nineteenth century, Americans became aware of the need to conserve natural resources, an impulse that began to take clearer form at the turn of the century. The end of the frontier was one source of anxiety, as the myth of superabundance that had guided American development met the reality of resource limits. So was the growing realization that much of the American West, arid and montane, would not accommodate to traditional agricultural settlement under the Homestead Act model. Much of what we call conservation grappled with that reality. The pace of industrial resource extraction also alarmed many observers, who worried not only about finite supplies of timber and other resources but also about the extraordinary waste and destruction they observed in processes of industrial resource extraction. And so conservationists of various stripes argued for the careful development and stewardship of natural resources, ideally by scientific experts employed by the federal government. Some conservationists were people we would today call preservationists, figures such as John Muir, who advocated for the preservation of areas of wild nature from the increasingly powerful forces of economic transformation. The creation of a series of national parks, monuments, and wildlife refuges between 1890 and 1920 testified to the power of this sentiment. Nonetheless, the dominant political strand of the movement in the years before World War I was utilitarian conservation, advocacy for the careful use of resources under the expert stewardship of resource professionals. The embodiment of this tradition was Gifford Pinchot, the nation's first professional forester and head of the U.S. Forest Service during the administration of Theodore Roosevelt. Pinchot famously argued for the wise use of federal forest resources based on a utilitarian philosophy that he hoped would guarantee a sustainable timber supply to serve the public good for the longest possible time. Water, grazing lands, wildlife, and minerals were also high on the conservation agenda. Soils, the nation's fundamental natural resource, seemed ripe for inclusion as well.[40]

This utilitarian tradition reached its apogee when President Roosevelt convened a Governors' Conference on the Conservation of Natural Resources in 1908. The main behind-the-scenes organizer of the conference was William John McGee, known to all as W J (he preferred not to use periods after his initials). McGee was a polymath, a self-professed expert in anthropology, ethnology, geography, geology, and hydrology. In the last decades of the nineteenth century he had worked with John Wesley Powell, first at the USGS and then at the Bureau of American Ethnology.[41] He was also a founder and past president of the National Geographic Society and the editor of its journal. The historian Samuel Hays called him the "chief theorist of the conservation movement," and Pinchot called McGee "the scientific brains of the conservation movement through all of its critical stages."[42] In 1907, just before taking on the organization of the Governors' Conference, McGee had switched positions yet again. That year, Milton Whitney had hired McGee to head a new unit on "Soil Erosion Investigations" within the Bureau of Soils. McGee's hiring seemed to represent a growing seriousness about erosion within the Bureau of Soils and a promising opportunity to make soils a part of the conservation agenda. Indeed, for a handful of conservationists, soil seemed ideally suited to the Progressive conservation framework, for here was a resource upon which all life depended, that took a long time to develop, and that Americans had spent down at a pace far more rapid than it could be naturally replenished. There also was a growing cadre of scientific soils experts ready to supply the apolitical expertise that Progressive conservationists thought necessary for the rational management of natural resources. McGee's arrival at the Bureau of Soils, then, seemed to suggest that soil conservation was poised to assume a central place in the Progressive conservation movement. Unfortunately, such hopes were premature.[43]

Soil conservation did get some play at the Governors' Conference, most notably in a speech by University of Chicago geologist Thomas Nelson Chamberlain titled "Soil Wastage." Years later, Bennett claimed that it was Chamberlain's speech that fixed his "determination to pursue that subject to some possible point of counteraction."[44] For his part, McGee did his best within the bureau to raise the profile of soil erosion and the need for conservation, most notably with his 1911 bulletin *Soil Erosion*, which the historian Douglas Helms has called "the Bureau's most complete treatment of the issue at that point."[45] But McGee died in 1912, and with his passing soil surveyors lost an important link to the larger conservation movement. Charles Van Hise, president of the University of Wisconsin and a proponent of resource conservation, gave soils and soil erosion some treatment in his seminal book, *The Conservation of Natural Resources* (1910). The Country Life Movement also directed attention to soils and the problem of soil erosion—the 1911 *Report of the Country Life Commission* included a section titled "Soil Depletion and Its Effects."[46] But in the end, soils were a secondary concern for most Progressive conservationists, whose interests centered on the nation's public land resources.[47] It would take a fuller sense of the scope of the soil erosion crisis and the next formative era in U.S. conservation history, Franklin Delano Roosevelt's New Deal, to give soils a central place in the conservation conversation and, more importantly, to create political space for conservation interventions on private agricultural lands. In the interim, what little soil conservation activity there was at the federal level occurred in the context of federal forestry, management of public grazing lands, and reclamation.

There was another important reason soil conservation struggled for traction during the Progressive Era: the theories and leadership of Milton Whitney. Even as he hired McGee to undertake investigations in soil erosion, Whitney was preparing another remarkable Bureau of Soils Bulletin that undermined the growing contention that soils needed to be conserved. *Soils of the United States* was the culmination of the Soil Survey's first decade of activity, and much of it was devoted to documenting the soils that the Survey had identified. But Whitney also reiterated his position that the nation's soils had plenty of nutrients and were in no danger of being degraded. "The soil is the one indestructible, immutable asset that the Nation possesses," Whitney insisted in *Soils of the United States*, which appeared in 1909. "It is the one resource that cannot be exhausted: that cannot be used up." Bennett allegedly reacted to this statement by noting, "I didn't know so much costly misinformation could be put into a single brief sentence." Whitney did recognize that soil erosion could be a problem. It could occur as one example of what he called "gross neglect." But his treatment of soil erosion in his 1909 report was nonetheless dismissive. He saw erosion as a temporary and localized problem; it was "but a detail making up the general average of national soil conditions." To those soil scientists and surveyors who had grown concerned about the need for soil conservation, *Soils of the United States* was a deflating report.[48]

That same year saw the release of the *Report of the National Conservation Commission*. Roosevelt had appointed the National Conservation Commission in the wake of the 1908 Governors' Conference, and its report was one of the most important documents of the Progressive conservation movement. Again Whitney's theories dominated the discussion on soils and agriculture. While the three-volume report did include several strong statements about the need to conserve soil, Whitney's lengthy essay "Crop Yield and Soil Composition" was entirely dismissive of soil exhaustion and did not even mention erosion. "The present practice of fertilization," he argued, "is based upon the fallacious doctrine that the soils become exhausted of plant food through continuous cropping and gradually become worn out." Again, he insisted that the "fertility of the soil is a great national asset which can not be exhausted."[49]

With Whitney in charge of the Bureau of Soils, soil conservation faced an uphill battle. Bennett attested to this in a relatively generous assessment of Whitney's leadership. "Professor Whitney," Bennett recalled in 1934, "usually wound up the discussions about getting something started [on soil erosion] by saying, 'Bennett, I don't think the time is quite ripe for getting into action with this problem.'" As a result, addressing the problem of soil erosion, Bennett recalled, "was delayed because the Professor was unable to get away from his principal lines of study to look seriously into the erosion problem."[50] As late as 1925, when Whitney published his opus, *Soil and Civilization*, he continued to dismiss soil erosion as an important topic. In a book that runs several hundred pages, he mentions erosion only once, as "a superficial action on the surface of the soil which will not be further discussed here."[51] Given the long history of books that have warned about the connections between soil erosion and the decline of civilizations, Whitney's absolute neglect of the topic is hard to fathom. Without leadership on soil conservation from the chief of the Bureau of Soils, those within the USDA who were becoming concerned about the environmental consequences of erosion were left to cobble together a picture of the problem and quietly evangelize.

The most important cobbler and evangelist was Hugh Hammond Bennett. In 1909, the same year that *Soils of the United States* appeared, Bennett assumed the position of "Inspector, Southern Division" for the Soil Survey, which meant that he was responsible for reviewing all of the county-level reports that came out of the region and correlating their results. It was an assignment that would be critical to his evolution as a soil conservationist. His wandering in the region became even more extensive. He visited most of the counties where soil surveys were in progress, and he developed a fuller sense of the region's soils and their degraded state. "Almost every soil survey in the Old Cotton Belt," Bennett recalled, "turned up startling information on the impoverishing effects of erosion."[52] A survey in which Bennett participated, of Lauderdale County, Mississippi (1910), revealed some remarkable gully erosion. Bennett and his coauthors warned of the area's volatile soil types and the need to practice careful conservation measures: "If the gentler slopes are not terraced and the steep situations kept in timber, deep gorgelike gullies or 'caves' gradually encroach upon cultivated fields, eventually bringing about a topographic situation too broken for other than patch cultivation."[53] These were some of the worst gullies Bennett had seen to date. His concerns would soon be heightened by two even more remarkable reports.

Few reports affected Bennett as deeply as the "Soil Survey of Fairfield County, South Carolina," completed in 1911. Fairfield County, which is just to the north of Columbia, the state capital, was at the heart of South Carolina's belt of lower Piedmont cotton lands in the nineteenth century, and it had suffered from grievous erosion. The 1911 survey identified ninety thousand acres in the county that had been cut to pieces by gullying.[54] The erosion in parts of Fairfield County was so bad that Earl Carr, who was in charge of the survey, struggled to complete his work. When Whitney grew frustrated with the slow pace of the survey and sent Bennett to investigate, Bennett quickly confirmed the extremity of the county's erosion problem. One surveying challenge was particularly important. The Fairfield County surveyors had initially mapped the worst gullied lands as "Caswell Sandy Loam," a soil type that other surveyors had invented in 1908 to make sense of some badly eroded bright leaf tobacco land in Caswell County, North Carolina.[55] As Bennett reported to Whitney, "In parts of the county general mapping is probably good enough because the county has been very badly dissected by erosion, in fact, one soil type, the Caswell, was designed to take care of these more severely washed areas."[56] But Whitney did not like this mapping. In his final edits of the report, he crossed out "Caswell Sandy Loam" wherever it appeared and replaced it with a phrase that was new to the Soil Survey lexicon: "Rough gullied land."[57] "The Rough gullied land," the final report noted, "comprises a variety of soils, which are so mixed, so badly washed and gullied, and so rough in topography that it was impracticable to make any satisfactory separation into distinct types." In previous surveys where erosion had been identified and mapped, the soil type was recognizable and could be rendered as a phase. But in portions of Fairfield County the problem was so dire that there was no way to identify eroded phases of particular soil types or to mark the boundaries between them. "Rough gullied land" was thus a classificatory throwing up of hands.[58]

The Fairfield survey also used another curious category. It identified forty-six thousand acres of "meadow," land that, despite this rather innocuous moniker, was almost as troubling to Bennett as the "rough gullied land." "Meadow" is a term that today invokes images of mountain grasslands filled with wildflowers, or landscapes such as Central Park's "Sheep Meadow," grassy swards designed to evoke the pastoral ideal. But historically the term has had a more specific meaning, describing low-lying grasslands that were seasonally wet or flooded and thus difficult to farm. In this precise usage, a meadow was a grassy wetland, as distinct from marshes, which are dominated by reeds and cattails, and swamps, which are generally forested. Surveyors thus used the term to describe soil types affected in their development by frequent inundation, and meadow soils represented another soil "phase." More importantly, though, soil surveyors working in the South mapped as "meadow" formerly fertile alluvial lands that had been substantially degraded by upland erosion. "Meadows" in this usage were areas where soil deposition had aggraded streambeds, resulting in much more frequent flooding, and where coarse sandy debris had buried productive alluvial soils. As the Fairfield survey put it, "Formerly the areas of Meadow were among the most productive soils of the county, but the wash from the uplands, which owing largely to poor management has been severe, has covered the original soils (Congaree) with deposits of variable texture, largely sand, and made many of them worthless for cultivation."[59] These forty-six thousand acres of "meadow" in Fairfield County were the yang to the yin of "rough gullied land."

Bennett was deeply moved by the Fairfield County report. It would be one of the cautionary tales he often invoked when preaching about the dangers of soil erosion in the South in the years to come. As importantly, he later referred to the Fairfield County soil survey as the "first erosion survey," by which he meant that it was the first soil survey in which lands transformed by erosion were mapped and quantified as such—in this case as "rough gullied land" and "meadow" rather than as distinct soil types or phases. It was the spark of an idea for mapping soil erosion that would be fully realized two decades later.

It was in the context of all of these various developments—the maturation of soil science, Milton Whitney's single-minded emphasis on soil texture and his founding of the Soil Survey to map it, the improvisatory evolution of soil classification and the anomalies produced by human-induced erosion, the growing concerns that some soil surveyors such as Hugh Bennett expressed about the need to conserve soils as part of a larger Progressive conservation movement, and a more general revolt against Whitney's theories by advocates of "permanent agriculture"—that David Long and his co-investigators descended upon Stewart County, Georgia, in 1912 to map that county's soils.

As in Fairfield County, the Stewart County survey provided a glimpse at a county wracked by spectacular gully erosion. David Long, the lead surveyor, was a prolific and experienced producer of county-level soil surveys for the state of Georgia, but he found the survey of Stewart County uniquely challenging. "This county is unlike any other county I know of in the state," he wrote to the chief of the Bureau of Soils, noting that many of its soils were new to him and that they were often difficult to discern because they were so intermixed by erosion.[60] As the survey progressed, Long's correspondence took on an increasingly frustrated tone. Not only was soil identification difficult, but traversing the county's broken terrain proved arduous and time consuming. By May 1913, Long let his frustrations boil over in another letter to Whitney. He was willing to take some of the blame for the delays and substandard results, but he also made it clear that the county's dynamic soils and topography had tried his patience. "Rather than drive over these worthless clay hills and rough broken land I thought it would be advisable to let the work stand and adjust it as best we can," he wrote. "The county is not worth mapping once much less twice."[61] He was ready to be done with the place.

Bennett first saw the soils of Stewart County on a weeklong inspection visit in February 1913 to investigate the causes of Long's growing frustration. In a letter to Whitney, Bennett was quick to praise Long, noting that he was "doing a splendid piece of work mapping out all of the areas of importance in considerable detail." But he was also clear that this was one of the most badly eroded places he had ever seen. He deemed the northern part of the county to be "as rough as any part of the entire Coastal Plain," comparing it to Lauderdale County in Mississippi and several badly eroded Alabama counties. But that was not the worst of it. "Outside of this rough northern portion of Stewart County," he wrote, "the area comprises a large total territory of Rough gullied land." There's no evidence that Providence Canyon, among the county's many gullies, particularly caught Bennett's eye during this trip, but over the course of a week he must have seen a fair portion of the county, so it's hard to imagine he missed it.[62]

When Long and his coauthors finally submitted their report in late 1913, they did make specific reference to the gullies near the Providence Methodist Church as extreme examples of the county's "Rough gullied land." More importantly, they sketched a broad picture of erosion in the county. According to the authors, 37,312 acres, or 12.5 percent of the county's land base, qualified as "Rough gullied land," which ranked it first in total area among all of the county's soil types. The second-ranking soil type, "Susquehanna clay," occupied another 12.4 percent of the county, and it consisted largely of clay subsoils on slopes that had been deeply impacted by the erosion of their surface layers. This was the northern part of the county that Bennett had referred to as the roughest piece of the Coastal Plain that he had ever seen. Finally, there were almost ten thousand acres of degraded "meadow" in the county, fewer than in Fairfield but still more than 3 percent of the county's land base.[63]

The authors of the "Soil Survey of Stewart County" included a photograph of "Providence 'Cave'" in their report. Like Veatch and Stephenson, they found it a compelling visual, though in their case it was an illustration "showing the character of rough gullied land." If they thought it picturesque, they did not say so. Toward the end of the report, as part of a page-long discussion of the various kinds of "Rough gullied land" found throughout the county, they again noted that "many of these caves are found in the vicinity of Providence Church." "One of the largest within the county," they wrote of Providence Canyon, "has developed in the memory of the present generation, having started with the formation of a small gully from the run-off of a barn.

The caves, some of which are 100 feet in depth and about 200 to 500 feet in width, ramify over large areas." The authors concluded that there was little possibility of restoring these lands to productivity, and they recommended establishing trees in the bottoms of the gullies as the best measure against their future encroachment on working agricultural lands. Beyond that, these spectacular chasms did not inspire a strong reformist message from the surveyors. But Bennett took note, and he soon made the gullies of Stewart County the centerpiece of his proselytizing.[64]

Despite the arrival of Bennett and the Soil Survey in Stewart County, Providence Canyon and the county's other large gullies only slowly entered the canon of degraded southern sites used by those writing about the problem of soil erosion. There were several reasons for this. First, the Stewart County report was not published until 1915, at which point the possibilities for highlighting the need for soil conservation within the Soil Survey had further constricted. Not only did Whitney continue to downplay the threat of erosion, but he and Curtis Marbut, who took over the day-to-day running of the Soil Survey in 1911, also pushed surveyors to back away from interpretation in their reports. Moreover, the Progressive conservation movement was beginning to lose steam as America moved toward war. Surprisingly, there was a small flurry of publications on soil erosion right around the time of the Stewart County survey's appearance: not only McGee's 1911 treatise but also two reports by Royal Davis, a former classmate of Bennett's at the University of North Carolina who worked with the "Laboratory Investigations" unit of the Bureau of Soils, titled "Economic Waste from Soil Erosion" (1913) and "Soil Erosion in the South" (1915).[65] In the latter report, which defined soil erosion as a particularly southern problem, Davis included photographs of what were almost certainly Stewart County's gullies.[66] But beyond Davis's work the gullies of Stewart County did not appear in the Progressive conservation literature.

By the early 1920s, Bennett began to use the Stewart County example, although obliquely at first, in his speeches and writings. More importantly, he began to use the logic of textural soil mapping to argue for removing certain soil types from agricultural production. In 1921, he published his first book, *The Soils and Agriculture of the Southern States*, which marked a culmination of his southern inspection work. The book mostly described the region's soil types and their particular qualities; there is surprisingly little discussion of erosion. But Bennett did mention the susceptibility of particular soil types to "ruinous erosion," and he described the "cañonlike hollows" that resulted when heavy rain caused "masses of soil to cave into the gullies from their heads and sides."[67] He may have been thinking of Stewart County when he wrote those lines. That same year, in a talk before the Third Southern Forestry Congress, Bennett indicated that there were areas of the Piedmont and Coastal Plain where soils could no longer support crop production and where timber production was likely a better economic use. Fairfield County was one of his examples, but so too was Stewart County. "The practical abandonment of approximately 25% of the area of a single county in West-central Georgia is a conspicuous example," he noted, of a place that ought to be planted in trees. By the early 1920s, then, Bennett was beginning to argue that particular soil types were prone to erosion and gullying, and that intelligent land utilization suggested that they be taken out of crop production and put into soil-conserving land uses.[68]

The period between Bennett's first visit to Stewart County in the early 1910s and the beginning of his aggressive proselytizing for a national program for soil conservation in the late 1920s was a peripatetic one for him. He was pulled in multiple, and often international, directions, which distracted him from publicizing the details of southern soil surveying and the problems of erosion. Not long after he assumed the position of inspector of all southern soil surveys in 1909, he went off to the Panama Canal Zone to survey the soils of this new American imperial territory and to assess its agricultural possibilities. Bennett's hagiographer, Wellington Brink, suggests that he might have been shipped to Panama to mute his growing agitation about southern soil erosion within the Bureau of Soils. Bennett also spent portions of 1914 and 1916 surveying the soils of Alaska. After a brief stint in the army during World War I that was cut short by two bouts of influenza, he spent a portion of 1919 as a member of a transboundary commission charged with settling a border dispute between Guatemala and Honduras. Then, in the early 1920s, Bennett traveled extensively in Mexico, Central America, and tropical South America as part of a Commerce Department assessment of the possibilities for rubber production in these regions. He also spent considerable time in Cuba on a survey of that island's soils, which resulted in his second book, *The Soils of Cuba* (1928). These international forays, combined with a 1920s conservatism that was generally hostile to conservation, meant that Bennett's evangelization for soil conservation was less successful than it would later be.[69]

Throughout this period, Milton Whitney's continuing reign at the Bureau of Soils also stunted Bennett's efforts to address soil erosion. Indeed, not until 1927 would the annual "Report of the Chief of the Bureau of Soils" even mention soil erosion as a major problem. It was no coincidence that the author of the report that year was A. G. McCall, who had just taken over for Whitney upon his retirement. McCall's characterization of the problem was quietly diagnostic of how Whitney had thwarted soil conservation efforts, and it suggested that Bennett had succeeded in getting his ear. "Another matter of immediate practical importance has forced itself to the front during the past year," McCall wrote:

> The tremendous loss from soil waste by erosion has been recognized in a rather hazy way for many years. The progress of soil-survey work in certain parts of the country has shown in increasingly definite form the magnitude of this loss. The loss through erosion has finally been recognized as a national menace, and a strong demand has been made for an investigation of the causes of the loss and for the working out of practical methods for its control. During the last year incidental attention has been directed to the matter, and the desirability, in fact the necessity, of giving the problem immediate attention has become manifest.[70]

The retirement of Milton Whitney in 1927 thus opened up substantial space for soil conservation concern at the federal level. Bennett would fill that space forcefully, and as he did, the gullies of Stewart County, and Providence Canyon in particular, rose to greater fame.

In the half-century between the Civil War and World War I, the gullies of Stewart County grew in relative anonymity, as did the historical conditions that would allow for their eventual discovery. When geological and soil surveyors arrived at them at about the same time, the early 1910s, their divergent tasks and worldviews meant that these scientific experts bestowed two differing sets of meanings on these gullies. The geologists, as we have seen, found something picturesque in Stewart County's deep gullies, an aesthetic judgment informed by a growing sense of the geological sublime then coming to be appreciated in the erosional landscapes of the West. The soil surveyors, on the other hand, were concerned with agricultural utility and the creation of a coherent and standardized body of knowledge about the nation's soils. While they were not the crusading soil conservationists that would rise to federal prominence two decades later, they did develop inchoate but growing concerns about the nature and extent of soil erosion as an agricultural problem, and they provided critical data for those few, Hugh Hammond Bennett preeminent among them, who were beginning to compile the "pertinent land facts" into an economic and moral case for soil conservation. But figures such as Bennett, as well as proponents of "permanent agriculture" such as Cyril Hopkins and Franklin Hiram King, had to contend with the unusual theories of Milton Whitney. Indeed, the unlikely influence of Milton Whitney is one the strangest parts of this story. While Whitney's Soil Survey, in its single-minded attention to soil texture, forced soil surveyors to contend with the extreme gully erosion in the South, his leadership and cornucopian theories about soil fertility thwarted the rise of a national soil conservation consciousness. Nonetheless, as the gullies of Stewart County continued to expand during the first quarter of the twentieth century, the attention these surveyors gave them set the table for a more open contest over their meaning in 1930s America.

PART TWO

Making Providence Canyon Meaningful in the 1930s

THE 1930S WERE the crucial years for making Providence Canyon meaningful. With the nation mired in an economic depression, many Americans became concerned about or experienced rural poverty and the incongruous problem of agricultural overproduction. For a nation clinging to the last vestiges of a Jeffersonian ideal of landed independence, this paradox seemed a perversion. As policy makers puzzled over these incongruities, their eyes invariably turned to the South and its exploitative systems of agricultural production. Famously, Franklin Delano Roosevelt labeled the South "the Nation's No. 1 economic problem," and the state of southern agriculture was a major reason for that prognosis.[1] The agricultural crisis of the 1930s was also an environmental crisis, as careless farming had wreaked havoc with much of the nation's land base. As the historian Sarah Phillips has written, "It is impossible to imagine the Great Depression in the United States without envisioning the era's environmental tragedies."[2] The Dust Bowl was the clearest of these tragedies, but soil experts also pointed to the South's tobacco and cotton areas, where erosion had been accelerating for decades. In no other region were the era's problems of economic underdevelopment and environmental degradation more tightly tied together. Providence Canyon rose to prominence in that context, during a broad reckoning with the social, economic, and environmental consequences of American, and particularly southern, agriculture.

But there were also countervailing forces that contributed to Providence Canyon's rise from anonymity in the 1930s. In the early twentieth century, Henry Ford was producing and marketing automobiles that middle-class Americans could afford, and the federal government made the fateful decision to join with state and county governments to modernize the nation's roads. As Americans adopted the automobile, one of their first impulses was to get back to nature and see the many natural wonders that the country had to offer. National parks and other scenic federal lands enjoyed huge increases in visitation and received infusions of federal money to improve their infrastructures to accommodate motor tourists. But state and local governments and booster groups also realized that unusual natural sites could be powerful tourist bait and that motor tourists could be useful stimulants to their economies. Federal funding for road building and recreational development was relatively high during the New Deal, when the Roosevelt administration put armies of unemployed people to work creating and improving state and national parks.[3] Just as the nation was becoming acutely soil conscious, then, many Americans also were eager to identify, develop, and visit new parks and other natural wonders.

The 1930s thus saw Providence Canyon take on two competing sets of meanings. On the one hand, its reputation grew as a scenic destination and natural curiosity, what many came to call Georgia's "Little Grand Canyon." On the other hand, it became an emblem of the environmental costs of southern plantation agriculture, "a spectacular testimony to man and his mistakes." Neither of these two interpretations would quite do justice to the place.

CHAPTER FOUR

A Land That Nature Built for Tourists

PROVIDENCE CANYON HAD ENJOYED a trickle of curious admirers during the early twentieth century. But in the 1930s the area experienced a tourist and publicity boom, largely as a result of a local and regional awakening to Providence Canyon's tourist potential. In 1933, for instance, the *Atlanta Constitution*, as part of an article and radio series on the "wonders of Georgia," profiled the "Caves of Providence in Stewart County" in a brief essay and full-page "Rotogravure Section" of photographs. While the captions recognized that the "remarkable canyon" had been created by "soil erosion due to careless farming practices," the paper seemed more intent on boosting the scenic aspects of the canyon. "It has attracted national attention, and is said," the paper added, "to be second only to the Grand Canyon in size and scenic beauty."[1] This is one of the earliest examples I have found of a comparison that, however specious, stuck to the place, marking it as a scenic wonder.

If any one factor encouraged locals to fix upon the tourist potential of Providence "Caves" when they did, it was the development of a series of paved through-highways that promised to bring more outside motorists to the area. East-to-West routes such as the Lincoln Highway had combined with various "See America First" movements during the 1910s to boost western motor tourism and visitation to the national parks, making travel to the region a nationalist imperative, particularly as World War I closed Europe to tourism. But this period also saw increased tourist interest in the South and efforts by southern boosters to build the roads necessary to bring tourists to their region. There were campaigns to create southern national parks in the Blue Ridge and Great Smoky Mountains and Florida's Everglades. Coastal areas saw increased resort development, as snowbirds flocked south seeking relief from northern winters. There was growing interest in the distinctive recreational hunting and fishing experiences to be had in the South, particularly in longleaf pine regions that had gained a reputation for salubrity. There was also a growing national fascination with a romanticized vision of the Old South that helped to fuel a tourist trade in the region. Together, automobiles, improved roads, a burgeoning interest in outdoor recreation, a growing affinity for warm climates, and a

{65}

sense among southerners that these developments could be profitable encouraged the residents of Stewart County and surrounding communities to offer up Providence Canyon as a tourist destination.[2]

The mother of all North-to-South tourist routes during this period was the Dixie Highway. First conceptualized in 1913 by Carl Fisher, an automobile headlight magnate and the motive force behind the Lincoln Highway, the Dixie Highway was more a network of routes than a single road. Completed by the late 1920s, it connected the upper Midwest with southern Florida, providing several routes through the Southeast. Fisher was a major figure in the Florida land boom and the father of Miami Beach, so the Dixie Highway served his real estate interests well. More broadly, though, the Dixie Highway and other roads like it promoted the interests of a group that historian Howard Preston has called "highway progressives," southerners for whom highway development, rather than improved farm-to-market roads, seemed to promise economic modernization for the region at a time when agriculture was in decline. "Highway progressives" were among the major boosters of Providence Canyon's tourist potential.[3]

For those communities left off the Dixie Highway's routes, agitation for additional through-highways in the 1930s was a common response. Heightened local and regional interest in Providence Canyon as a potential tourist site was directly related to such highway projects. The most important was the Florida Short Route, which originated in Duluth, Minnesota. The Florida Short Route passed into Georgia from Alabama at Columbus and headed south through Albany before terminating in Lake City, Florida. The main route pulled motorists through Webster County, just to the east of Stewart County. In July 1935, as the *Atlanta Constitution* again featured "Providence Caves" in one of its Sunday photo spreads, a large motorcade celebrated the completion of the last paved section of the Florida Short Route in Georgia. The *Stewart-Webster Journal* used the occasion to editorialize about the importance of the road.[4] At that same moment, some of the state's leading citizens organized a "See Georgia First" movement, anchored in the newly organized Georgia First Association, and they invited civic organizations and municipal governments from all over the state to send information about sites in their communities that might have tourist appeal.[5] The possibilities for capitalizing on motor tourism, which were robust even in the midst of the Depression, led Georgians to think anew about their surroundings. As Nelson Shipp of the *Columbus Enquirer* noted in a speech to the Lumpkin Lions Club in 1937: "Thousands of people in the North are looking for places to go—new places to visit, the unique in the nation. These see-America-first people are turning more and more to the South, and are discovering our region rapidly. The development of the automobile and the paved road bids fair to increase this as time goes by."⁶

The growing conviction that paved roads and automobile tourism could be a boon for the region grew in uneasy proximity to the arrival of New Deal conservation. Even as the local newspaper celebrated the completion of the Florida Short Route and the rise of the "See Georgia First" movement, it also noted the creation of a Civilian Conservation Corps (CCC) camp in Stewart County, just four miles south of Lumpkin. The camp's two hundred men and boys were to undertake erosion control projects in cooperation with the newly created Soil Conservation Service. The paper noted, without apparent irony, that "the boys will be located in a close proximity to the famous Providence Caves, which stand a little less in National Prominence to the Grand Canyons [sic]."⁷ The CCC camp opened in July 1935, housing more than two hundred. Even as there were early stirrings of an effort to promote Providence Canyon as a tourist site, then, the county was also confronting its erosion problems and taking advantage of federal conservation programs that delivered labor and expertise.

While local leaders generally welcomed New Deal conservation to the area, the campaign to boost Providence Canyon as a tourist destination emerged partly as a result of locals becoming defensive about their county's growing reputation for substantial soil erosion. The *Atlanta Constitution* again featured "The Grand Canyons of Stewart County" in a September 1935 pictorial spread, but this time as a place "Where the Federal Government Will Battle Nature to Preserve Homes and Farm Lands." In an accompanying article, the paper announced the coming of the Soil Conservation Service, with an army of CCC labor, to Georgia to combat its serious erosion problems. "Camp No. 9 in Stewart County," the article noted, "is the one which will attempt the conquest of Providence cave, probably the largest man-made gully in the humid regions of the United States—a grisly monument to the destructive power of erosion and a grim testimonial to the folly of land misuse."⁸ A headline in a September 1935 issue of the *Stewart-Webster Journal*, "Stewart's Famous Caves Illustrated in Government Reports to Show Results of Neglected Lands," suggested that locals were aware of how soil conservationists viewed the county's gullies.⁹

By the spring of 1937, several prominent locals, fed up with the growing prominence of "Providence Caves" and other Stewart County gullies in the national conservation literature, defended their county and began to build a case for the beauty and naturalness of these landscapes.¹⁰ Soon the *Stewart-Webster Journal* offered its own willful reinterpretation in a short piece titled "Stewart County's Famous Caves Scenes of Natural Beauty Rather than Horrors of Destructive Elements." "During the past several months," the article noted, "critics have seemingly found and seen only the argument of soil erosion in the Providence Caves of Stewart County, while, in other days, those who love scenic beauty and the wonders of nature have driven scores of miles from all directions to see these beautiful pictures of nature as no other place except the Grand Canyons [sic] have portrayed the earth's stratas of soil, the sea level of eons ago, and the beauties of the sunset from these dear old hills of Stewart."[11] More and more, the region's residents, buoyed by the possibilities of lucrative tourist traffic, would champion and build an aesthetic case for what they called Providence Caves, a case, as the above quote suggests, that relied upon a geological sublime.

During the autumn of 1937, the *Columbus Enquirer* initiated an editorial campaign to have Providence Canyon designated as a national park, an effort led by their editorial writer, Nelson Shipp. Various local and regional civic organizations quickly fell in line. For a couple of years the campaign was vibrant, attracting substantial attention to "Providence Caves" and producing a dramatic spike in visitation. Initially, and somewhat confusingly, Shipp and the *Enquirer*'s editorial board expressed an interest in protecting Providence Canyon as both a state *and* a national park, but as their campaign gained steam, national park status became their preferred goal. Throughout the process, they focused on two qualities that they thought defined Providence Canyon. First, they insisted that it was a spectacle of national importance, comparable with the great landscapes of erosion in the American West that had already become national parks. Second, they insisted that these spectacular "caves" were natural.

In October 1937 the *Enquirer* officially launched its campaign to make "Providence Caves" a national park. In their first editorial on the subject, the editors provided clarification on this local appellation, sensing that it might confuse outsiders. "Not really caves but gorges," the paper noted, "they received their name as such because they were created by the clay beneath the topsoil caving in and being washed out of the apertures, which have increased to major size until the 'caves' now constitute a 'Little Grand Canyon.'"[12] The term "caves" was thus meant to indicate that persistent cave-ins had produced the deep gullies, or, as the boosters preferred, "gorges." Despite this explanation, many found the moniker misleading. As a result, at a meeting of the Lions Club of Lumpkin, Reverend J. N. Shell of the Lumpkin Methodist Church suggested a name change: in all of their future resolutions in support of the park idea, they would refer to the place not as "Providence Caves" but "Providence Canyons." The *Enquirer* picked up on the name change quickly.

An editorial published a few days after the meeting noted that the term "caves" gave "the public the wrong idea and as a result it is not able to anticipate the colorful grandeur of the walls in the open sunlight."[13] The name change stuck. It's not clear when the moniker became singular rather than plural, but it's likely that the dissonances between the plural "Providence Canyons" and the singular "Little Grand Canyon" had something to do with it. Semantics clearly mattered in the packaging of a tourist site.[14]

The most striking aspect of the efforts to make Providence Canyon into a national park was how consistently proponents premised their campaign on the proposition that the canyons were natural formations. In their opening editorial on the subject, the *Enquirer* editors praised the "caves" as a "strange creation of natural forces," "a wonderful natural phenomenon," and "one of the true show places in mother nature's garden of the earth." In another editorial a few days later they exclaimed: "When Nature scooped out the acres now missing from the spot and washed them from the location, she had a definite eye for beauty."[15] The Lumpkin Lions Club, the self-fashioned progressive organization of Stewart County professionals and business leaders who constituted the main force of the campaign, called the area an "unusual and awesome natural wonder of great scenic beauty" in their resolution of support for the park idea.[16] The editors of the *Stewart-Webster Journal* were no less insistent in their editorializing. "The hand of Nature has been both artisan and artist in going about the creation of this alluring spot," they wrote of what they called a "charming creation of Nature." "Manifest Destiny seems well to have intended the construction of a natural wonder," they prophesied. "Indeed, no better name could have been chosen for it than 'Providence Canyons.'"[17]

Local park boosters almost never mentioned that humans contributed to the creation of the spectacle, even though they clearly understood that others saw these gullies as disastrous human creations. But there was one curious exception to this rule, a theory that had legs among the early boosters but then faded. In his first presentation to the Lumpkin Lions Club, Nelson Shipp gave a talk titled "From Indian Trails to Little Grand Canyon," in which he suggested that the gullies had first developed long ago as a result of water eating away at Indian trade routes down to the Chattahoochee River. As the local paper reported it, Shipp "discussed the theory of one of the governmental authorities that the canyon started with the Indian trails of long ago, that through the years the crevices were made in the woodlands along these trails of the red men by the rainfall, until finally great apertures existed." But those Indian gullies, according to Shipp, "were healed over or covered up with trees and vegetation, until the day came, about seventy years ago, when suddenly erosion started again and re-created the canyons." "Of course, this is only a theory," Shipp admitted, but its provenance, having come from an unnamed "expert in the soils field," gave it some credence in his mind.[18]

There are several points worth noting about the Indian trails theory. First, I found no substantiation of the theory or even any information on who the alleged "expert in the soils field" might have been who first proffered the theory. Second, those who entertained the theory blaming Indian trails effectively recognized human culpability in one era and among one group but scrupulously avoided doing so for their own era. The *Stewart-Webster Journal*'s convenient use of the passive voice—"suddenly erosion started again"—was bad enough, but just a few lines below its reporting of Shipp's theory, the paper referred to "the hill into which Nature is eating." Whether there is any validity to this Indian trails theory or not, it is remarkable that locals offered it as speculation without any corresponding recognition that they and their ancestors might have had something to do with the creation of Providence Canyon. Third, the Indian trail story is one of a number of origin myths that have had lasting power, even in the absence of substantiation. When locals got a historical marker erected to help direct tourists to Providence Canyon, it read in part, "Trickles of water running down old Indian paths to springs formed the Providence Canyons, natural wonders of the Southeast."[19] Helen Eliza Terrill, in her compendious *History of Stewart County, Georgia*, repeated this theory, noting that Indians traveling to a trading area at Florence had worn trails into the natural terraces, which gullied as rainwater channeled into them.[20] As we will see, this was only one of several intriguing and perhaps diversionary theories about the origins of gully formation in Stewart County.

Along with their insistence that Providence Canyon was a work of nature, another major theme of the park campaign was the desire to "swell the tourist crop." This emphasis on how important national park designation would be to capturing tourist dollars was ubiquitous and unabashed. Such sentiment was not only characteristic of local interests eager to create national parks in their areas so as to attract tourist dollars; it was particularly true in the South, where boosters often looked jealously at the West and its profusion of national parks and insisted on getting equal federal treatment. Park-making entered the realm of pork barrel politics during the interwar period, and the local campaign's efforts to get federal recognition was partly about the dollars that would come with it. Providence Canyon, the *Columbus Enquirer* noted in a succinct summary of its editorial position, was "a land that Nature built for tourists."[21]

Setting aside Providence Canyon as a national park was about more than crass commercialism, however. As spectacular natural areas became lodestones for the tourist trade during the automobile era, preservationists constantly found themselves swimming against the commercial tide, attempting to protect nature from the billboards, souvenir shops, and hawkers of bed and board that materialized along the nation's roadsides. Americans increasingly saw national parks as havens from commercial tourist development, even as national parks spurred such growth on their outskirts. It would be a stretch to suggest that the local proponents of national park status for Providence Canyon were selfless preservationists putting the interests of nature above profit. But local boosters did recognize the power that national park status would confer on Providence Canyon. The label "national park" suggested a limited number of authentic natural areas of *national* interest. Any area that achieved such status would join a pantheon of tourist sites. In contrast to the private, commercialized roadside natural wonders, the "rock cities" and "mystery spots" that competed for tourist traffic by making incredible claims about their uniqueness, national parks were certified and trustworthy. Their status was all the advertising they needed. Not surprisingly, locals salivated at the prospect of having Providence Canyon discussed in the same breath as Yellowstone, Yosemite, and the Grand Canyon. National park status for Providence Canyon promised to both swell the tourist crop and raise the canyon above the commercialism that was reshaping so much of the American roadside during this era. A national park protecting a great "natural" wonder would be a source of local, state, and regional pride. It would put Stewart County on the map.

Local support in Stewart County soon coalesced around the *Enquirer*'s proposal. Supporting the Lions Club in its endorsement of the idea were the mayors of Lumpkin and Richland, the county's two major towns, and the editor of the *Stewart-Webster Journal*. The county schools superintendent, the county agricultural extension agent, the local Daughters of the American Revolution chapter, and a number of other prominent citizens and organizations also lent their support. Not surprisingly, the *Enquirer* reported, "A. F. Kelly, owner and proprietor of an automobile agency in Richland, thinks the proposal would be a splendid development."[22] Regional support was also strong. James Woodruff, president of the Chattahoochee Valley Chamber of Commerce, got behind the plan, as did the Columbus Chamber of Commerce. Even H. L. Manley, the superintendent of the local CCC camp, supported the park idea.[23] Here were Preston's "highway progressives" in full array.

All of this support and the spurt of publicity generated by the park proposal amplified visitation. The *Enquirer* reported that more than one hundred automobiles visited the canyons on Sunday, November 7, 1937, and that, "with the increase in travel, gasoline stations, drug stores, and lunch stands are ex-

Figure 9. Colorized postcard of "Providence Canyons." Undated but likely from the late 1940s or later. Collection of the author.

periencing an increased business."[24] A week later, the *Enquirer* reported giddily, more than "three hundred automobiles containing some fifteen hundred people visited Providence Canyons" on a single Sunday. By the end of November 1937, Mrs. F. W. Ellis, whose family owned some of the collapsing land around Providence Canyon, was keeping a registration book at a small visitor stand the family ran on the edge of the canyons, a book that would record visitors for the next three decades. While the register mostly documented visitation from the immediate region, there were tourists from quite a number of other states as well as a few dignitaries. General George Patton and his wife came from Fort Benning for a visit, accompanied by Mrs. Henry Stimson, wife of the secretary of war. There also were souvenirs and hand-painted postcards of Providence Canyons available for purchase.[25] Visitation would continue at these elevated levels through the end of the decade.

As the park campaign accelerated, comparisons with the iconic erosional landscapes of the West became more common even as they stretched credulity. By the time the *Enquirer*'s campaign began, for instance, the "canyons" were also known as the "Royal Gorge of the Chattahoochee," a disingenuous moniker since the Chattahoochee had nothing to do with carving them. The reference nonetheless placed the site in the company of the Royal Gorge along the Arkansas River near Cañon City, Colorado, which was also known as the "Grand Canyon of the Arkansas."²⁶ "The 'Royal Gorge of the Arkansas' is not one-tenth as beautiful" as Providence Canyon, the *Enquirer* boasted, although it failed to note that Providence Canyon was only about one-tenth as deep and nowhere near as long. Another common point of comparison was the Garden of the Gods, an impressive collection of sandstone formations just outside of Colorado Springs, Colorado. "'The Garden of the Gods' in Colorado Springs is a marvelous thing, and people have gone thousands of miles to see it," the *Enquirer* editors opined, "but it does not compare with Providence Canyons." Another common comparison was to Utah's Bryce Canyon, which became a national monument and then a national park in the 1920s. At one point Nelson Shipp of the *Enquirer* referred to Providence Canyons as "Brice [*sic*] Canyons of the East." At the very least, the local boosters insisted, there "is no natural wonder just like the Stewart [C]ounty Canyons east of the Rocky Mountains."²⁷

The comparisons between Georgia's "Little Grand Canyon" and the Grand Canyon proper also became more elaborate. Not only did boosters insist that Providence Canyon was comparable to the Grand Canyon, a group of locals who served as guides gave names to the various formations within Providence Canyon that echoed those given at the Grand Canyon. The paper reported that formations within the canyons had been given such names as Cleopatra's Throne, Puritan Village, Ghost's Wall, Hall of Ghosts, National Cathedral, Lion of Lucerne, Venetian Lagoon, Temple of the Shrines, Blade of Damocles, Little Switzerland, Sleepy Hollow, Spirit Canyon, Neptune's Banquet Seat, and Taj Mahal.²⁸ The *Enquirer* even claimed that while the Grand Canyon was "tremendously large, awesome, and colorful," it had neither "the gentle tints nor the intriguing formations that Providence Canyons possess."²⁹ Local supporters were shameless in their hyperbole, but the goal was to convince visitors and federal officials that Providence Canyon deserved national park status.

As visitation grew and Providence Canyon took on new significance as a tourist site, local park boosters began to exert political pressure. A Lions Club resolution called upon political representatives from the area to throw their support behind the plan, something that U.S. Congressman Steve Pace seemed eager to do.³⁰ Pace visited Providence Canyon on November 8, 1937, and he promised to take action when he returned to Washington.³¹ A week later, he, along with Georgia senators Richard Russell and Walter George, sent letters to the Columbus Chamber of Commerce expressing their support for the project.³² Russell and George also wrote to Robert Fechner, head of the Civilian Conservation Corps, to ask if the CCC would pitch in on the park project. Fechner responded by assuring the park supporters that, were the land made into a national park, the CCC would consider lending their support in developing the area for visitors. But Fechner also noted that "the area in question is privately owned and under existing legislation it would not be possible to undertake this improvement and development." Senator Walter George then wrote a follow-up letter to the Columbus Chamber of Commerce, urging them to pursue state legislation that would allow for public purchase of the necessary lands.[33] A couple of weeks later, Pace wired local supporters asking that they send him pictures of Providence Canyon so that he could make his case in Washington. But the *Enquirer* also sounded an ominous note about creating the park: "This however hinges upon the definite assurance that the lands there can be obtained at a reasonable price. It is not believed that the government would be interested in paying much for the property, as it classes such as eroded lands and not worth much to the owners anyway."[34] The federal government had already showed substantial willingness to buy up "submarginal lands." But, as the *Enquirer* intimated, the government, whether federal or state, was not likely to pay a premium for land that was badly eroded, no matter how scenic.

These concerns aside, the New Deal context gave local boosters great hopes for achieving their goals. Franklin Roosevelt was a frequent visitor to the region, and many thought he would take an interest in the proposal. Roosevelt's "Little White House" in Warm Springs, Georgia, was about eighty miles to the north. Local boosters hoped Providence Canyon might become part of a larger tourist itinerary that would include stops at Warm Springs, where Roosevelt had recuperated from polio in the 1920s, and Pine Mountain, where Roosevelt had a 2,700-acre farm and where CCC workers were creating what became F. D. Roosevelt State Park, Georgia's largest.[35] The *Enquirer* suggested that Roosevelt be invited to visit Providence Canyon on his next trip to Warm Springs, and the Lumpkin Lions Club included a presidential invitation in one of its resolutions. "The President has discussed the phenomenon," the *Enquirer* editors wrote of the county's gullies, "with one of the government's soil experts, and is believed to be deeply interested in it *from more than one standpoint.*"[36] This final phrase was an oblique recognition that there was more than one way of interpreting this landscape, a point the paper had made with unusual frankness in an early editorial on the national park proposal:

> The opportunity seems to be a tremendous one to turn what some have regarded as a liability into a large asset. The Caves are securing considerable publicity over the nation already, as a natural wonder and curiosity. Manifestly, it would be an excellent thing to take advantage of this wide publicity—"cash in" on it, so to speak—and turn the trend of public notice into one of tourist-

bidding. Henceforth, the publicity would be guided into majoring upon the natural wonder and beauty of Georgia's Little Grand Canyon, instead of having it principally a discussion of erosion and how many acres of agricultural land and their wealth have been swept away and lost to production.[37]

The *Enquirer* assiduously followed its own advice during the last few months of 1937, pushing both the beauty and the tourist potential of the area without mentioning the human role in producing this spectacle.

Local boosters finally succeeded in getting the attention of national and state park officials in 1938. The Columbus Chamber of Commerce invited Arthur Demaray, assistant director of the National Park Service, to visit Providence Canyon and assess its viability as a national park.[38] While Demaray declined, the Park Service sent two other officials to visit Providence Canyon and assess its park worthiness: Roy E. Appleman, a regional Park Service historian from Richmond, Virginia, and B. C. Yates, who was in charge of the Kennesaw Mountain National Battlefield Park, a Civil War site outside of Atlanta.[39] By late April 1938, Appleman had issued a report. He praised Providence Canyon's "considerable scenic attractions" and noted that he had "not seen anything comparable to it in coloration in the eastern United States." The area reminded him, he wrote, of Bryce Canyon. But he ultimately recommended that Providence Canyon would be more suitable as a state park than a national park. He also argued for "a gigantic project of erosion control" to stabilize the canyon walls so that roads and visitor facilities could be built without fear that the expanding canyons would swallow them up. Appleman's report appears to have killed local hopes for national park designation.[40]

As the national park idea faded, Georgia's interest in making Providence Canyon a state park grew. Local park boosters hosted a delegation of state park and conservation officials, including B. F. Burch, head of the State Department of Conservation, and Charles M. Elliot, state director of parks and recreation.[41] Several weeks after his initial visit, Burch returned to address members of the Lumpkin Lions Club, telling them that he thought they had the makings of a first-class state park. Then, in late August 1938, Governor E. D. Rivers visited Providence Canyon to inspect the site and give a speech on its rim. Rivers called the site "a remarkable piece of natural scenery that should be developed," and he insisted that the state would move on the park idea as soon as the residents of Stewart County purchased the necessary land and turned it over to the state.[42] Local boosters believed that if they could raise the money to purchase the land, either from county coffers or from private donations, the benefits would more than compensate for the costs. But the project languished. In 1939, federal authorities relocated the CCC camp from Stewart

County to Perry, Georgia, much to the dismay of local residents.[43] There went the labor necessary to develop tourist facilities. Of more consequence, local park boosters had a hard time raising funds for the purchase of the necessary lands. While they lauded the benefits of park creation and praised the "progressive citizens of the county [who] realize this golden opportunity," they also pointed out that "others do not want to recognize the real value to our county that a state park could and should be."[44] As World War II approached, even the state park campaign was spinning its wheels.

A final disappointment for this campaign came with the 1940 publication of *Georgia: The WPA Guide to Its Towns and Countryside*. As its subtitle indicated, *Georgia* was the product of the Works Progress Administration (WPA) and its Federal Writers' Project, one in a series of guides on each of the forty-eight states plus several cities and territories. As the local park boosters paged through their copies of *Georgia*, their initial reaction must have been one of pride, for "Providence Caverns" made an appearance in the guide's first photo insert, juxtaposed with images of other iconic Georgia landscapes such as Brasstown Bald, the highest point in the state. But as they read further, anger replaced pride. As in most of the other state guides, the second half of *Georgia* consisted of a series of "tours" through the state. The goal of these itineraries was to highlight the tourist offerings in each state and to encourage tourism as a road to economic recovery. In a sense, this was a federal realization of the "See Georgia First" effort that had first encouraged boosters to package Providence Canyon as a tourist site. But the way that the WPA writers portrayed Providence Canyon left local park supporters crestfallen.[45]

In narrating a tour from Columbus south to the Georgia-Florida line, *Georgia* highlighted the ravages of erosion in the surrounding countryside. Just to the south of Columbus, the guide pointed out to potential travelers, "the land has been seriously eroded, which has caused great suffering not only on the farms but in the towns. The ravages sometimes show in spectacular canyon formations." Once the tour arrived in Lumpkin, the authors were quick to note the particularly egregious erosion of Stewart County, citing the figure, derived from the 1913 Stewart County soil survey, of forty thousand acres of land "rendered unproductive by erosion." Of the houses and farms along the county's roads, the guide painted a stereotypical picture of southern rural poverty: "Lean hounds sleep in the shade of the open passageways, and chickens scratch in the bare, sandy yards. Red clay gullies, bordered by pine trees and broom sedge, stretch from the road back far into the fields." It was a description suggestive of *Tobacco Road*, Erskine Caldwell's brutal portrait of a Depression-era Piedmont farmscape near Augusta, Georgia, not the prose of a tourist brochure. A detour to Providence Canyon followed. "Although the gullies are a spectacle of destruction," the guide tersely noted of what it called "Providence Caverns," "their great size and the delicate colors of their vertical walls give them a strange beauty." This was a grudging admission that the "caverns" were worth seeing, though it was hardly a gushing endorsement of Providence Canyon as a spectacular natural feature.[46]

As local supporters struggled to push forward a state park proposal that was losing steam in the early 1940s, the editors of the *Stewart-Webster Journal* rued the content and tone of the newly released WPA guide. In an August 1940 editorial titled "As Others See Us," they reprinted the entire offending section from the guide. "Our first reaction is to get indignant and yell 'sabotage,'" the editors admitted, and they protested that the authors had opted to "see the weaker points of our land and civilization." But they also wondered whether they had really put their best foot forward, and they used this bad publicity to urge local residents to work doubly hard to improve the impressions that outsiders might garner of the county. They were engaged, the editors intimated, in a fight to frame how others saw them, and if this was the best they could get from a tourist guide, they knew they had more work to do. They needed to get outsiders to appreciate Providence Canyon as a natural wonder, not as a human blunder.

The local boosters and civic leaders who pushed for park preservation were the main street elite of town life in interwar Georgia. They represented a narrow slice of Stewart County society. Newspaper coverage from the era gives little sense of how the majority of the county's white population, many of whom were still working the land, felt about the proposal. Perhaps, behind the overweening New South enthusiasm of the town and county leaders, there was skepticism among others of the county's white farm families that slowed the park proposal. The inability to purchase the land for park purposes suggests as much. But if their opinion is difficult to gauge today, it is not nearly as elusive as the perspectives of the largely disenfranchised African Americans who, by 1940, constituted almost three-quarters of the county's population. So it was with great enthusiasm that I greeted an indelicate 1941 headline in the *Stewart-Webster Journal*: "Darkey of Stewart Author of Poem Book 'Canyons at Providence.'" The brief article noted that Thomas Jefferson Flanagan, "a native born Stewart Countian, born and reared in the hills of Stewart," had recently produced a book of poems inspired by Providence Canyon.[47] Here was an African American perspective on the place, one that, like the WPA guide, offered a more ambivalent interpretation.

Thomas Jefferson Flanagan was born in Stewart County in 1890 and grew up in the western part of the county, not far from Providence Canyon. At thirteen, he began sending off samples of his poetry to various publications

Figure 10. Thomas Jefferson Flanagan. This image of Flanagan appeared in *The Canyons at Providence*.

in Atlanta, where his work caught the eye of Frank L. Stanton, then serving as the first poet laureate of Georgia. With Stanton's help, Flanagan was soon getting his poetry published in the *Atlanta Constitution*, and success at other publications followed. Flanagan received a B.A. from Atlanta University, a historically black college (which merged with Clark College in the late 1980s to become Clark Atlanta University), and he went on to a career as a teacher and principal, an editor, and a U.S. railway mail clerk. He authored several other books of poetry and a novel of plantation life, *Jimson Weeds*, establishing a minor literary reputation. But it was his slim volume of poetry, *The Canyons at Providence (The Lay of the Clay Minstrel)*, that caught my attention.[48] The poem's curious parenthetical subtitle betrayed the influence of Sir Walter Scott and his early romantic poem, *The Lay of the Last Minstrel* (1805). Scott's writings were so popular among antebellum southerners for their romantic renderings of chivalrous knights that Mark Twain famously blamed the Civil War on "Sir Walter disease" and its delusionary effects. It seemed an odd model for an African American poet to be using. I had to wonder whether Flanagan was being ironic or even playing on how white Americans had redefined minstrelsy as a racialized form of entertainment in the nineteenth century.[49]

The *Stewart-Webster Journal*'s brief announcement of the publication of Flanagan's ode to Providence Canyon was positive and prideful. The article gave no intimation that Flanagan's interpretation strayed from the newspaper's scenic boosterism. Flanagan's brief foreword to *The Canyons of Providence* suggested that his would be a similarly celebratory and naturalistic approach to Providence Canyon. He effused about the "hanging gardens and varied clay colored tapestries of old Providence Canyons—said to be second only to the Grand Canyons of the Colorado," and he made explicit his desire to add the canyons to the list of renowned southeastern literary landscapes, invoking it alongside "The Marshes of Glynn," the eponymous subject of Sidney Lanier's ode to coastal Glynn County, Georgia, and the "palmy bays of Florida."[50] But the poem itself is decidedly ambivalent, and it does not fit easily with the booster perspective.

"'Tis mine to find beauty where tragedies spread," Flanagan insisted in his "Prelude," comparing the deep gullies to the "pits of my soul, / Eroded by the tears I have wept for my clan." Flanagan clearly recognized not only the tragic nature of the erosion that had produced such a stunning place but also an African American story carved into these hills, a story of the heroic "forgotten man," as he refers to his race, and his sorrows and hopes:

> But, as high as the clouds in their rambling round
> Are the hopes that he harbored, while shackled to the ground
> That loosed not its rivets, 'till dissolving like snow
> It took to the rivers at the Great Furrow's blow!
> As the dwarfed beaten trees kneel low on the crags
> I bow where before knelt his courage in rags.

This mixture of sorrow and hope, of suffering and courage, of bondage and dignity was what Flanagan found as he contemplated the canyons at Providence.[51]

The main body of the poem, which Flanagan titled "The Erosion March" to invoke not only the travels of the gully heads but also the transience of the region's black tenant farmers, begins with a similar effort to square "beauty and tragedy." Referring to the gullies, he wrote:

> Oh, these restless travelers—have you seen them?
> Opening like volumes through the blue pine lane,
> Where the exiled strugglers starved between them
> On what was left of the cotton and the grain.
> But a strangeness of beauty and tragedy blends them
> And though here's the boneyard of the farmer and his farm,
> The art-thirsty eye, admiringly defends them
> For their romantic glory and rerety [sic] of charm.[52]

For Flanagan, the beauty of these formations and the tragic circumstances of their creation were not merely the logical makings of an irony but essential parts of Providence Canyon's poetry.

In his poem, Flanagan referenced several of the stories and myths of canyon formation, but he turned them into different sorts of object lessons. Invoking the "immortal penman" that produced the canyons, and thus suggesting divine intervention at work, he wrote, "From one creaking barn-eave came your lust for muse." But he also continually insisted that there was human agency at work in the canyons. "The master of the lyre is come and troubles he the clay," Flanagan wrote of the natural or divine forces he saw at work in Stewart County. But he followed immediately with a stanza invoking human design: "Where lowly sons of poverty charted him a way."[53] He also wrote of the graves that fell into the chasms and of the road that, with each year, traversed a narrower and narrower ridge between expanding gully systems. He saw in these workings of providence a literal and symbolic leveling, a sense of justice in the race-blind way the land collapsed in western Stewart County, taking the graves of both black and white former residents:

> The road was the dividing line betwixt the black and white
> Placed in the last encampment ere this leveling blight
> Contending that all dust is dust and there is neither rank
> Nor creed beneath the challenge of the master of the flank:
> Hamp Mayo died a chattel but fond freedom knows his ghost
> And sawing on his fiddle he surveys the haunting host
> Cap'n McGinty, whose estate was all the lower ranch
> In disgust claps his palsied hands at each avalanche.[54]

Ultimately, despite the sorrows and injustices to be found in the canyon, Flanagan saw it as a symbol of forgiveness and glory, the beautiful product of an unlovely era:

> But here is erased all their blame, God forgave and hence
> While their folly deep is writ, here's a recompense—
> Unmarred by biting chronicles of broken bone and sweat
> We inherit a coliseum no mortal hand has set.

Despite the many ways in which he saw history written on the canyon's walls, then, Flanagan ultimately refused to see these gullies as merely the work of humanity. They truly were the canyons of providence, a gift of beauty left to memorialize the tragic history that produced them.

There is a millennial optimism in Flanagan's poem, a sense that Providence Canyon represented not only a scar symbolic of racial injustice but also a beautiful and timeless gift of healing and forgiveness, what he called a "masterpiece prophetic that will march on with the sun!" We cannot, of course, infer from this one poem any broad and singular African American perspective on Providence Canyon, but Flanagan's is an intriguing interpretation, one that reveals some of the silences at the heart of the local elites' efforts to see only "nature" at work. Thomas Jefferson Flanagan saw in Providence Canyon a prophecy, a stone of hope in a gully of despair.[55]

Just as Flanagan's poem appeared in print, World War II seems to have sapped the remaining energy from the local park movement. What scant press coverage Providence Canyon received in the years after the war tended to focus on how little known the place had become. In 1954, Bernice McCullar, a writer from the *Christian Science Monitor*, visited Providence Canyon and was shocked that the "remarkable canyon has never been commercialized." She even lamented how poorly marked the extraordinary spectacle was. "At the turning of a little country lane down in Stewart County, Ga., there is an amazing sight," she wrote: "A gigantic gully larger than any other this side of the Grand Canyon. Yet there are no signs that bid the tourist not to miss it. No painted lures along the tourist's path admonish him to include it in his sight-seeing. You have to ask the way to it; you are likely to know it only by chance."[56] A year later, *Collier's* had a brief article on "Georgia's Little Grand Canyon" that struck a similar note. "Down a side road off U.S. Route 27 in Stewart County, Georgia, is one of the most spectacular natural wonders in the United States," wrote Bern Keating. "Yet tourists rarely see it, for it is little publicized."[57]

Providence Canyon finally became a state park in 1971, after several decades of local pleas to state legislators.[58] Hugh Carter of Plains, Georgia—Jimmy Carter's cousin, his successor in the Georgia Senate, and a worm farmer and purveyor of live bait—sponsored the enabling legislation, which hinged on the successful purchase of the land in and around Providence Canyon. By that point, the bulk of the land for the park, about 750 acres, belonged to the Ingram and LeGrand Lumber Company, while several local families owned the remainder. State funds paid for much of the land, though funding also came from the U.S. Bureau of Outdoor Recreation, a short-lived federal agency that emerged out of the Outdoor Recreational Resource Review Commission created by Congress in 1962 to survey outdoor recreational opportunities and to work with states and localities to create more parks. Land purchase was complete by the end of 1970, and Providence Canyon State Park opened on July 1, 1971, not long after Jimmy Carter became governor. Many of the legislative sponsors of the park predicted that it would prove a great success. Hugh Carter prophesied that Providence Canyon would turn into the state's biggest tourist attraction. In another homage to the Grand Canyon proper, there were even plans for mule-train tours down into the bottom of the canyon. But these visions never came to pass. Over the last several decades, Providence Canyon State Park has only managed to be a local and regional tourist attraction at best.[59]

The 1930s saw a vibrant campaign to preserve Providence Canyon as a natural wonder, a land that nature built for tourists. Celebrating it as a grand work of nature comparable to the spectacular erosional landscapes of the West, local and regional boosters tried to make Providence Canyon into a park, first a national park and then, when that failed, a state park. At times their boosterism was shameless, and their insistence that Providence Canyon was a work of nature sticks in the craw. Nonetheless, such a campaign was of a piece with the sorts of tourist development occurring throughout the South and the nation, as local chambers of commerce and other "progressive" citizens sought to capture a burgeoning tourist trade by advertising spectacular natural areas and gaining for them the official imprimatur of park status. Turning Providence Canyon into Georgia's "Little Grand Canyon" served those purposes well. But that claim to fame would be challenged by divergent interpretations of Providence Canyon, as the ambivalent renderings of the WPA guide and Thomas Jefferson Flanagan both intimated. Ultimately these local boosters lost the public battle to interpret Providence Canyon in the 1930s.

CHAPTER FIVE

Giving Fame and Focus to the Fact of Soil Erosion

IT WAS ANOTHER GROUP that more successfully made Providence Canyon famous in the 1930s. For soil conservationists, New Dealers, environmental writers, and even a few southern liberals, Providence Canyon served as the poster child of southern soil abuse. I first stumbled upon this reformist interpretation of Providence Canyon while I was searching through the Library of Congress online collection of Farm Security Administration photographs. Beginning in 1935, under the leadership of Roy Stryker, the photographic section of the Resettlement Administration (which became the Farm Security Administration in 1937) employed some of the nation's best photographers to fan out across America and document the social and environmental conditions they encountered. The FSA roster included Walker Evans, Dorothea Lange, Russell Lee, Ben Shahn, Jack Delano, Marion Post Wolcott, Gordon Parks, and Arthur Rothstein. I wondered whether any of these photographers had trained their lenses on Providence Canyon. After a quick search, I hit pay dirt: a series of nineteen photographs of Providence Canyon and other big gullies, many of which were attributed to Arthur Rothstein and titled simply "Erosion, Stewart County, Georgia." The others were unattributed and untitled, but my guess is that Rothstein took them as well. A selection of these images appears as the opening gallery in this book.[1] I knew about Rothstein and his documentary photography primarily through his famous Dust Bowl photographs, which had done so much to sear that tragedy into American consciousness. Providence Canyon had been captured by one of the key visual artists of 1930s America, and in his hands the canyons seemed to illustrate a set of profound environmental and regional problems. Coming across these Rothstein photographs was a revelatory moment, for it confirmed a hunch and triggered a broader search. If Rothstein photographed these gullies, surely others had also fixed upon them as well.[2]

The 1930s were years of tremendous environmental concern, much of it rooted in a poignant sense that Americans, over several centuries of agricultural expansion and resource exploitation, had pushed the land too far. Where

{83}

the Progressive conservation story that emerged a generation earlier was one of impending resource limits, the New Deal narrative was about environmental overshoot. New Dealers spoke of a nation that had not just come up against scarcity but one that had transgressed its limits as a result of greed, misplaced optimism, and poor planning. In contending with environmental overshoot, New Deal conservationists often turned their attention from public land resources to the soils of private farmlands. In fact, a signature feature of the New Deal's "New Conservation" was its strong correlation between soil erosion and rural poverty.[3] Out of this concern came a flurry of "soil jeremiads," moral tales anchored in a historical narrative of frontier exploitation.[4] In my search for Providence Canyon's former fame, I thus turned to these and related writings, hoping to find a few references to the place. What I found was the extensive deployment of Providence Canyon as an object lesson, a stark depiction of the sins at the dramatic core of the New Deal's environmental morality tale. The New Deal was the era of the land scar, and few scars were more dramatic than those to be found in Stewart County, Georgia.

The towering figure, the Jeremiah, in making soil matter to Americans during the New Deal years was, of course, Hugh Hammond Bennett. Bennett had been building a case for the seriousness of soil erosion, particularly in the American South, for several decades, but in the late 1920s, after a period of relative quiet on the subject, he began, in his own words, "to howl about the evils of soil erosion."[5] In 1927, as Milton Whitney finally ceded the leadership of the Bureau of Soils and as a massive Mississippi River flood prompted questions about the condition of America's watersheds, Bennett initiated a campaign for a national program of soil conservation.[6] In these years before the drama of the Dust Bowl, he often featured the gullies of Stewart County in his campaign. In an April 1927 article in *Scientific American*, for instance, Bennett warned about Americans' profligate destruction of the soil, and he cited a "county of the coastal-plain region" where sixty thousand acres had been ruined by extreme erosion.[7] In early 1928, Bennett delivered a series of lectures at the USDA Graduate School that focused on soils and soil erosion, and he referred to Stewart County as the place where "some of the worst gullying that has taken place on cultivated land anywhere in the United States is exhibited." He specifically cited the "the Providence Cave neighborhood" as the most extreme gullying to be found in the county.[8] That same year, in an essay that appeared in the prominent academic journal *Geographical Review*, Bennett again discussed the gullies of Stewart County and provided a dramatic photograph of Providence Canyon as a feature illustration.[9] And in a 1929 *North American Review* article titled "Our Vanishing Farm Lands," Bennett cited the case of Stewart County yet again.[10] Over the next several years, Bennett constantly invoked the gullies of Stewart County to dramatize the need for a national program of soil conservation.

Bennett's most important publicity volley in his effort to make soil conservation a national priority came in 1928, in a USDA booklet titled *Soil Erosion, a National Menace*, which he cowrote with W. R. Chapline, the U.S. Forest Service's inspector of grazing.[11] Historians have seen *Soil Erosion, a National Menace* as the foundational document in Bennett's campaign for a national soil conservation program. In part 1 of the report (Chapline wrote part 2, which was focused on western grazing lands), Bennett meticulously documented the long, slow processes of soil creation and how rapidly poor farming practices could undo them. Then he listed several parts of the country where soil erosion had proceeded to such a degree that the prospects for future agriculture were all but ruined. One was Fairfield County, South Carolina, whose 1911 soil survey had so shocked him.[12] Bennett also mentioned the "loessal region" of Mississippi, where severe erosion had also occurred. It too had become a staple of his writings and speeches. But the most striking photo in the report was of one of Stewart County's gullies, and in the text of *Soil Erosion, a National Menace* he discussed Stewart County's erosion at length. A "county in the Atlantic Coastal Plain has 70,000 acres of former good farm soil, which, since clearing and cultivation, has been gullied beyond repair," he observed. "In one place where a schoolhouse stood 40 years ago gullies having a depth of 100 feet or more are now found, and these finger through hundreds of acres of land, whose reclamation would baffle human ingenuity."[13] As Bennett built a case for a national soil conservation program, Providence Canyon featured prominently.

After a campaign of just a couple of years, Bennett succeeded in convincing Congress and the U.S. Department of Agriculture to take soil erosion seriously. At his urging, and with the assistance of A. B. Connor of the Texas Agricultural Experiment Station and Representative James Buchanan of Texas, Congress appropriated $160,000 for the USDA to undertake soil erosion investigations beginning in 1930. The USDA used the funds to set up ten soil erosion experiment stations, and Bennett became the director of its Soil Erosion and Moisture Conservation Investigations.[14] While the various erosion investigation stations set up around the country studied the mechanics of erosion in different soil types, bringing important data to a problem that had only been studied anecdotally, Bennett also pushed the idea of comprehensive erosion surveys. The New Deal would eventually make that vision a reality.

In September 1932, in the depths of the Depression and just two months before the transformative national election that would usher in the New Deal, Bennett embarked upon a tour of the South, where he took in the sites of extreme soil erosion that he had first encountered during his days as inspector of the southern soil surveys. He revisited Fairfield County, where he saw nothing but "[d]estroyed land, worn out and abandoned as far as the eye could reach." He then headed to Stewart County, which he had last visited in 1913. "I found on this second trip that the gullies had not stopped with their chiseling away of the fine agricultural lands," he wrote. "They had grown longer, deeper, wider; they had branched out, forming new canyons." "Providence Cave," he noted, "is one of these enormous gullies, probably the deepest man-made gully in the western hemisphere."[15] A local farmer named B. E. Howell, who had lived in the area for decades and recalled attending the school that had collapsed into the gullies decades earlier, toured Bennett around the vicinity.[16] "Countless ravines have gutted the landscape that formerly was a beautiful undulating countryside, unscarred and clothed with rich sandy loam soil," Bennett concluded. "Wretched methods of land use have whittled away the substance of the earth, leaving in the wake a desert-like waste."[17] This reunion tour clearly refreshed in his mind these spectacular examples of erosion's ruinous impacts, for he would moralize about them with increasing fervor over the next decade.

The coming of the New Deal further elevated the cause of soil conservation and Bennett's prominent role within it. Franklin Roosevelt and many of his most important advisors, from Rexford Tugwell of the USDA to John Collier of the Bureau of Indian Affairs, saw soil conservation as a critical initiative and incorporated it into their various programs. In July 1933, the Special Board on Public Works allotted $5 million for soil erosion prevention. After some bureaucratic intrigue, Henry Wallace, the secretary of agriculture, offered Bennett the directorship of the newly created Soil Erosion Service (SES), a work relief program centered on erosion prevention. Because concerns persisted about the legality of federal soil conservation work on private lands, the SES, which was housed in the Department of the Interior, focused on a series of demonstration projects spread around the nation. While funding for soil conservation was improving, this was not quite the comprehensive soil conservation program Bennett had hoped for.[18]

While Bennett was getting the Soil Erosion Service up and running, he also served as a member of the Land Planning Committee of the president's National Resources Board (NRB). In that capacity he realized another of his major goals: a comprehensive survey of soil erosion across the entire country. Created in the summer of 1934, the NRB was tasked with preparing a comprehensive report on the nation's land and water use, a report completed by December 1934 and submitted to Congress early in 1935. Few documents better express the rage for land planning that characterized New Deal conservation policy. The report argued that the nation's farm problems were partly rooted in the cultivation of marginal farmlands, many of which were susceptible to erosion, and it recommended that the federal government purchase seventy-five million acres of such lands and retire them from production. In the process of preparing the NRB report, the Soil Erosion Service completed the Reconnaissance Erosion Survey of the United States, a rapid national survey of erosion conditions carried out by 115 soil erosion experts who visited almost every county in the United States between August and October 1934. The most important result of that erosion survey was a map, one similar to those produced by the Soil Survey but that charted erosion conditions rather than soil types. A black-and-white version of that map, with eight different erosion categories, appeared in the NRB's December 1934 report (see figure 11). The following year, in a supplemental report titled *Soil Erosion: A Critical Problem in American Agriculture*, there appeared the full-color foldout version of the Reconnaissance Erosion Survey of the United States map, which featured thirty different soil erosion categories. This survey and the map that resulted from it was, for Bennett, the culmination of three decades of soil surveying and the logical extension of the erosion mapping that had begun haphazardly in the soil surveys of Fairfield County, South Carolina, and Stewart County, Georgia. The direst of the thirty erosion categories on the map was "Destroyed by Gullying," examples of which appeared in blood-red shading. Not surprisingly, two of the most prominent red stains betrayed the locations of Fairfield and Stewart Counties.[19]

As the Reconnaissance Erosion Survey was coming to fruition, the SES was at a crossroads and Bennett's priorities were shifting. Through its brief history, the SES had bounced around between various relief agencies and faced the important limitation of not being able to work on private lands. Its funding was scheduled to expire in June 1935, and the future of federal soil conservation was uncertain. Meanwhile, Bennett's geographical focus was shifting. While his expansive interest in soil erosion had pulled him out of the South and into the West by the late 1920s and early 1930s, he had mostly focused on two western subregions during the early New Deal: the Navajo Reservation, where overgrazing seemed to be contributing to erosion, and the western cotton regions of Oklahoma and Texas, which had developed their own substantial erosion problems.[20] But by the spring and summer of 1935, as the SES seemed to be in its waning days, the southern Great Plains became the nation's exemplar of the perils of soil erosion.

The first substantial dust storms to develop on the Great Plains had a powerful influence on both Bennett and national soil conservation politics and policy. On March 20 and 21, 1935, as Bennett testified before the House Public Lands Committee about a bill to create a permanent federal soil conser-

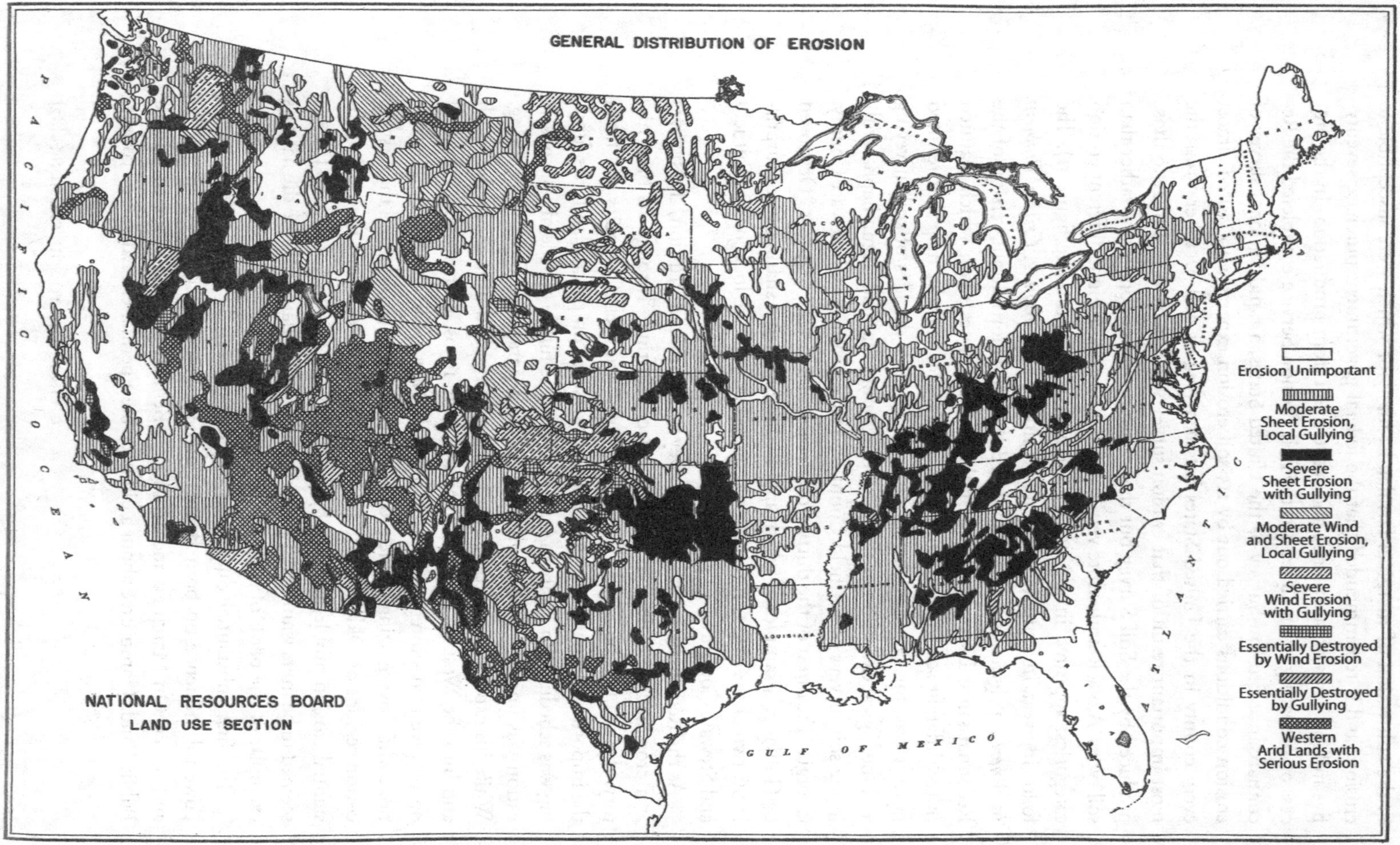

Figure II. "General Distribution of Erosion." Map produced by the Soil Erosion of the United States. This map appeared in the *National Resources Board Report*

vation agency, one of the worst dust storms of the season was building force, carrying tons of topsoil eastward from the plains. As legend has it, Bennett stalled the committee by poring through and placing into the record the data from the Reconnaissance Erosion Survey. Then, as the dust storm arrived (and off the record, so the accuracy of the story from here on out is less reliable), Bennett allegedly got the committee members to leave their seats and peer out the windows as dust darkened the skies of the nation's capital. "Everything went nicely thereafter," Bennett later recalled of the hearings. Bennett's biographer referred to this as "the turning point" in Bennett's campaign to create the Soil Conservation Service and in his rise to national conservation prominence. It's a good story, though it likely makes this serendipitous moment of environmental collusion more of a pivot point than it was.[21] Nonetheless, the Public Lands Committee did issue a favorable report on the bill on March 29, and the full House approved it on April 1. As the bill moved over to the Senate, the dust storms on the plains intensified. April 14, 1935, came to be known as "Black Sunday" because of a fierce dust storm that darkened the southern plains that day. That evening in the *Washington Evening Star*, a reporter named Robert Geiger first referred to the southern plains as a "dust bowl." The next day, on April 15, the Senate passed the Soil Conservation Act, and after one last trip to the House, the bill arrived at the White House soon thereafter. President Roosevelt signed it into law on April 27, 1935.[22]

Bennett, who became the first director of the new Soil Conservation Service, finally had his comprehensive national soil conservation program, thanks in large part to the Dust Bowl. "Without the Dust Bowl's potential for theater," Donald Worster has concluded, "there most likely would not have been such a large commitment of money and federal personnel to protect the soil." This was certainly true, but such a conclusion risks obscuring behind a cloud of dust the extent to which federal institutions for charting and countering the nation's soil erosion problems already had made substantial progress prior to the decisive dust storms of 1934 and 1935.[23] As influential as the Dust Bowl may have been in the end game that led to the creation of the Soil Conservation Service, we would be mistaken to assume that soil conservation began with the Dust Bowl. The cause had been slowly building strength for more than a quarter of a century, and it had been the South's gullies that had provided the early theatrics.

Throughout the mid-1930s, even as the Dust Bowl came to dominate soil erosion consciousness, Bennett continued to raise the specter of Providence Canyon. In a June 1934 feature article in the *New York Times*, Bennett insisted, "Stewart County in Georgia affords an excellent illustration of the insidious destructiveness of man-induced erosion."[24] Later that year, in another essay in

Scientific American, Bennett juxtaposed the spectacle of the accelerating wind erosion on the Great Plains with the "major evil" of water erosion, using the gullying in Stewart County as his sole example of the latter phenomenon: "The 70,000 acres of land destroyed by erosion in this single Georgia county represent but a fraction of the gigantic stride America has made in the direction of land misuse and consequent land impoverishment and destruction."[25] And in 1938, in the annual USDA Yearbook of Agriculture, which that year was titled *Soils and Men*, Bennett and Walter Lowdermilk included a picture of a gully head in Stewart County and a brief discussion of gullying in the county.[26] Bennett clearly had not forgotten the region whose erosion problems had radicalized him.

Bennett's magnum opus, *Soil Conservation*, appeared in 1939, and at almost one thousand pages it is an exhaustive (and exhausting) treatment of the topic. But one only has to make it to page four before the gullies of Stewart County make an appearance, as figure 2. Figure 1, on page 2, is a photo of "virgin forest," its aboriginal soils anchored and intact. This first image served as the contrapuntal setup to Stewart's massive chasms, the first visual representation of extreme soil erosion that came to Bennett's mind. The caption for the photo, a striking aerial shot of cultivated land giving way to steep badlands, reads: "This maze of gullies, which crosses an entire county in the Southern Coastal Plain and cuts through portions of two adjoining counties, has permanently destroyed more than a hundred thousand acres of some of the best land in the locality. The original gully began about 60 years ago in a cultivated field. Stewart County, Georgia."[27] Later, when he published a much shorter text, *Elements of Soil Conservation*, Bennett used the same two images to start the book, though this time he made the gullies of Stewart County into figure 1.[28] Bennett saw these gullies as illustrative of the erosion problem he was doggedly trying to get Americans to take seriously. What better way to shock them into action than to show the most extreme case he could think of. That meant Stewart County.

One of Bennett's main goals in publicizing soil erosion was to get Americans to recognize the problem not just as a series of localized and isolated natural disasters, but as a persistent, widespread, and often hidden scourge of agricultural land use. Soil erosion, he wrote hopefully, "is understood now not as a freak of nature which occasionally turns up on isolated farms and ranches, but as an almost continually active process which attacks countless fields, whole watersheds, and broad farming communities."[29] Providence Canyon and Stewart's other gullies were attention getting, but, in their extremity, they may not have been ideal illustrations for making that larger point. Bennett admitted as much later in his book, providing another photograph of the

Figure 12. Gullies, Stewart County, Georgia, figure 2, from Hugh Hammond Bennett's *Soil Conservation* (1939). This network of gullies still exists today just to the northeast of Providence Canyon State Park.

gullies as well. "What has happened in Stewart County, Georgia," he wrote, "is a striking example of the extremes to which erosion can sometimes progress when land is farmed without proper precautions."[30] Bennett realized that Providence Canyon and the other gullies of Stewart County were outliers, but they were striking visuals and so he kept returning to them.

Another New Deal environmental writer critical to Providence Canyon's rising fame was Russell Lord. Lord was one of the most popular and prolific writers about the American land in the 1930s and 1940s. He was also a founder of the pioneering environmental organization, Friends of the Land, and served as the editor of *The Land*, its journal. More than any other organization of the era, Friends of the Land—which was founded in 1940 and counted among its members such luminaries as Aldo Leopold, Paul Sears, Rexford Tugwell, Morris Cooke, and Hugh Hammond Bennett—made "permanent agriculture" its rallying cry.[31] Lord began his career as a journalist who wrote about agriculture, and he then took a job with the USDA during the early days of the New Deal, where he wrote (and ghost-wrote) articles and speeches on New Deal agricultural policy. In 1934, Lord left the USDA and Washington,

D.C., and moved with his wife Kate to a twenty-two-acre farm in rural Maryland, embracing the tenets of the back to the land movement. A year later, in 1935, Bennett, by then the head of the Soil Conservation Service, contacted Lord about whether he would be interested in writing a report on the soil erosion problem in the United States. Lord accepted the assignment, traveling on and off for three years throughout the country to see firsthand the extent of the nation's erosion problems. From this assignment came not only the government report that he had been commissioned to write, *To Hold This Soil*, but also a spin-off book, *Behold This Land*, both of which appeared in 1938.[32]

Providence Canyon made prominent appearances in both publications. In *Behold This Land*, there was a photograph of "Providence Cave" in the first glossy insert, and Lord devoted several pages to a discussion of the county's gullies. Lord began his discussion with a sturdy piece of canyon lore, one that Bennett had already helped to popularize: that Providence Canyon had started as a result of a persistently dripping barn roof. This was the most prominent of a series of local creation stories about the county's gullies. When local journalists and other visitors to the area enquired about the causes of extreme gullying in the county, they heard other stories as well. We have already seen that some locals entertained the idea that Indian trails down to the Chattahoochee had resulted in substantial gullying. One journalist later reported the theory of a local named Ida Ward, who believed that "the canyons started from trickles of water running down paths that Indians and wild animals used to get to springs of water. She rejects the popular legend that water dripping from the eaves of a barn cut the original gash in the clay surface."[33] Another story had it that gullying had started as a result of cattle repeatedly using a particular path to get down to a spring.[34] Yet another story, this one particularly fantastic, suggested that one of the gullies was the result of a local woman repeatedly, and apparently with great force, dumping her wash water in the same place, which ate away at the soil and initiated gullying. Such stories, which had little to do with poor agricultural practices or the larger political economy of the plantation South, raise some interesting questions about the relationship between human land use and these gullies, questions to which I will return to later. But Lord was not interested in such stories. Most important to his purposes was to describe the voraciousness with which these gullies had been swallowing the county's farmland. "Providence Cave" was a cancer, he insisted, metastasizing unopposed.[35]

Like other "soil jeremiads," *To Hold This Soil*, the official government report, was a plea to care for the nation's soil after several centuries of sinful soil mining, and Lord, who had an apt name for an author in the genre, again featured images and discussions of some of the worst eroded and gullied land that the nation had to offer. Here too he included a lengthy discussion of "Providence Cave" as well as a full-page photograph. Again he saw sinister forces at work. He described the "fingers" of the gully as "malign living structures, creeping, feeding on soil, on the habitations of the living, on the bodies of the dead." "In 1913, when fieldmen of the Bureau of Soils making the first American soil maps reached Stewart County," Lord continued, "they found a great part of it devastated." As Lord made clear, soil scientists had discovered Providence Canyon several decades earlier. But it was not until the 1930s, when soil conservation became a celebrated cause, that the erosion problems of Stewart County achieved national attention.[36]

Providence Canyon appeared in numerous other conservation publications during this era. In his 1936 text *Soil Erosion and Its Control*, Quincy C. Ayres, an agricultural engineering professor at Iowa State College, featured an image and brief discussion of Providence Canyon.[37] In *Soil Erosion Control* (1937), Austin Earle Burges noted, within the first few pages of his book, that what he called "Providence Gully" was "known to all students of soil erosion for its immensity" and was "probably the deepest man-made gully in the Western Hemisphere." In a caption under two photographs of Stewart County gullying, Burges lamented: "Ten thousand acres, including farms, roads, woods, barns, homes, and churches, were swallowed up by this gully—yet a simple terrace or diversion ditch built at the right time could have stopped it all."[38] There is a photograph of the canyon in William van Dersal's book, *The American Land* (1943), which, though it does not identify Providence Canyon by name, notes: "This chasm is almost 200 feet deep. It was caused by man and was started by unwise farming methods."[39] That's about as economical a summary as one will find of the lesson that this crowd of interwar conservationists hoped to teach by bringing Providence Canyon before the nation's eyes. A few years later, in *The Land Renewed: The Story of Soil Conservation*, a short book that van Dersal coauthored with Edward Graham, Providence Canyon was featured in photograph and text in a brief chapter titled "Big Gullies." And Karl B. Mickey used Providence Canyon as the supreme example of gully erosion in his book *Man and the Soil* (1945). In a caption beneath a photo of Providence Canyon, he scolded: "This gully in Georgia is one of hundreds of similar examples of erosion caused by human negligence."[40]

International observers also took note of Providence Canyon as an exemplar of accelerated erosion. In 1936, the British geographer F. Grave Morris toured the Southeast for several weeks in the company of the legendary American geographer Carl Sauer and a group of SCS officials, with the goal of examining the erosion problems of the region. Along with his former student and SCS employee C. Warren Thornthwaite, Sauer was then involved in organizing a series of erosion research projects at various sites throughout the United States.[41] The year following their tour, Morris published an essay in the *Geographical Journal*, a British publication, titled "Soil Erosion in South-Eastern United States." He had seen a number of large gullies throughout the region, but he focused on Stewart County's. "Another remarkable gully," he wrote, "occurs in the soils of the coastal plain near Americus, Georgia, known locally as the 'Grand Canyon.' This must be more than a half-mile in length and . . . it must be 150 to 200 feet in depth and perhaps 300 feet wide." Morris drew a set of larger lessons from Providence Canyon and its ilk. "It is impossible to do justice to the appalling results of soil erosion in the south-eastern states," he exclaimed. "An exploitative and wasteful system of agriculture has ruined the soil and this, in turn, is reflected in the terrible conditions under which the majority of the farmers, both white and negro, are living."[42]

On the international stage, U.S. soil scientists had been pioneers in bringing attention to the problem of soil erosion in the 1920s, but the British were not far behind, though they generally studied the phenomenon in relation to their colonial possessions. In the late 1920s, Britain's Colonial Office created the Imperial Bureau of Soil Science, a clearinghouse for information on soil erosion and conservation efforts in the colonies. In the introduction to their sweeping and amply illustrated study *Vanishing Lands: A World History of Soil Erosion*, G. V. Jacks, who was then the director of the Imperial Bureau, and his coauthor, R. O. Whyte, described the various types and stages of erosion until they got to the most extreme, gullying. As their example of this process, they turned to Providence Canyon, explaining how a dripping barn roof became a rill and then a rivulet and then "a torrent that tore away the soil and subsoil over an ever-widening area and flung whole farmsteads into the gaping wound." Their point was that erosion had to be stopped in its early stages or the results could be extreme. That they chose Providence Canyon as their example suggests its increasing global renown in the years before World War II.[43]

Back home, the Soil Conservation Service began making profligate use of images of the gullies of Stewart County to accompany its publications. Providence Canyon made a prominent showing in the second issue of *Soil Conservation*, the journal of the SCS. The author was Leon J. Sisk, an official at the SCS Sandy Creek Demonstration Project in Athens, Georgia. Sisk's article, titled "All of This Started from the Trickle from a Roof," attributed the canyon to runoff from the roof of a barn constructed by a farmer named Patterson in 1855. After providing these few factual tidbits, Sisk moved into a more imaginative realm, invoking the Pattersons' barn as a symbol of the richness that he imagined had marked the family's agricultural life prior to the soil neglect that degraded so much of the county. The barn, he speculated, invoking a symbol of permanence, must have been full of stored grain and feed for the farmer's livestock. But then the rains came, the land began to melt away, and farmer Patterson neglected the rills and gullies that developed, letting his prosperity and security wash away. As a narrative starting point, this vision of settled abundance allowed Sisk to hammer home the soil conservation gospel and its chief lesson: that we were turning good land into bad through our inattention. "The largest gully in Georgia," he concluded, "which has caused thousands of dollars damage, which would cost thousands more to stabilize, could have been prevented."[44] Just two years later, *Soil Conservation* again featured the gullies of Stewart County, this time in an article written by Wellington Brink, titled "Some Cracking Big Gullies." Brink also worried that the gullies were swallowing valuable farmland in the region, "bale-to-the-acre land, much of it." "It's a spectacular scenery," he admitted, perhaps aware of the building movement to boost Providence Canyon as a tourist site. "But it's a scenery that Stewart County could do better without." "The gullies of Stewart County," Brink concluded, in a nice description of the role they had come to serve, "give fame and focus to the fact of soil erosion."[45]

One of the most honest and provocative confrontations with Providence Canyon during the 1930s came from the pen of Jonathan W. Daniels, the editor of the *Raleigh News and Observer*. Daniels was the scion of a famous North Carolina family. His grandfather had been governor of North Carolina. His father, Josephus Daniels, founded the *News and Observer* in 1894 and had infamously put it in the service of the Democratic Party's politics of white supremacy as the decade ended, working to defeat the Populist political challenge and playing no small role in fomenting the brutal 1898 Wilmington Race Riot. Josephus Daniels went on to serve as secretary of the navy under Woodrow Wilson, a position he held throughout World War I, and he later served as the U.S. ambassador to Mexico during the New Deal years.[46] On the issue of race relations, Jonathan, who took over the editorship of the *News and Observer* in 1933, was among a group of white southern liberals who rose to prominence in the 1930s and who were amenable to dismantling Jim Crow. He would go on to briefly serve as press secretary for both Franklin Roosevelt and Harry Truman.[47]

During the summer of 1937, Jonathan Daniels embarked on a sweeping tour of the South in preparation for his book *A Southerner Discovers the South*, which appeared in 1938. The book is most notable as a watershed in Daniels's evolution as a racial progressive, but he also devoted attention to the southern landscape. In chapter 31, titled "Graveyard and Gully," he recalled rising before daybreak one morning so that he would have time for a side trip, recommended to him by his friend and fellow North Carolinian Hugh Hammond Bennett, during his long drive from Atlanta to Tallahassee. "You should by all means see the famous Providence Cave in Stewart County, Georgia, near the town of Lumpkin," Daniels remembered Bennett imploring him: "This is a celebrated gully probably more than 150 feet deep at the head, yet formed in soil within the past half century. It is but one of numerous similar gullies which have ruined a large area of good land in Stewart and two adjacent counties."[48]

As a southern progressive at once critical of and sympathetic to his region and its ills, Daniels was ambivalent about Providence Canyon when he finally came upon it. Daniels pulled into Lumpkin on a Sunday morning, driving past the "starched children, black and white" who were on their way to Sunday school. Getting directions for the "caves," he headed west. "They are, of course, not caves at all," he reported. "They are ditches. But ditches of the same genus as the grand canyon of the Colorado. Down through the red soil to almost pure white clays the chasms run in the midst of cultivated Georgia farms. They come perilously close to the highway and seem ready to engulf road and farm-house and church." While lying on the edge of one such gully, Daniels wondered at the "points of earth ris[ing] like clay towers of stalagmites." They seemed to him "fine phallic symbols in the midst of an advancing sterility."[49] Few quotes captured the place and the 1930s battles over its meaning quite so well.

Daniels noted that such gullies were nothing new to Georgia. He mentioned the famous Lyell Gully and reprinted the description that Lyell had offered of it a century earlier. Staring down into Providence Canyon from its edge, Daniels's thoughts then turned to the land use practices that had produced "what we ordinary folk in these times call erosion," and to the lessons that flowed there from. He noted that the decline of cheap western lands had forced southerners to take erosion seriously, and he admitted that erosion and soil exhaustion were disproportionately southern problems. This was a major lesson to be gleaned by any serious and open-minded southerner attempting to discover the South in the 1930s.[50]

But Daniels also made a perplexing observation about the gullies of Stewart County, one that Bennett's description had not prepared him to grapple with. "They left me not so much shocked at the land destroyed," he wrote of the canyons, "as puzzled by the character of the destroying ditches." "The Stewart ditches grew from careless man and washing water," he freely admitted. "But the land about them did not seem to an unpracticed eye badly worn, soils about to collapse in dramatic canyoning. Instead, the big ditches lay deep and open in the midst of apparently rich and fertile fields. Corn, cotton, and pecan trees grow near them. And near them, too, I came upon a big barn smelling of animal husbandry. There were houses, too, surrounded by such flowers as are generally grown in the country only by such folks as love the earth as well as hope to profit from it."[51] The gullies were dramatic, and it was hard not to see in them the region's broader history of debilitating erosion. But there was also something that did not compute about the juxtaposition of fine working lands and deep gullies. As the day progressed, more visitors arrived at the "caves," surrounding Daniels on the canyon's rim. This was just months before the *Columbus Enquirer* started its campaign to create a park, and Sunday afternoons were the most popular time to visit. As he watched the small crowd gather, Daniels concluded that "on that Sunday afternoon the rived land did not look tragic; the people upon it seemed not at all wasting survivors or wasted remnants of a human order that had departed from depleted and eroded soil. Far otherwise, land and people—white folk and Negroes—seemed to me that day vigorous and arrayed for pleasuring."[52] Daniels was not so sure that the gullies of Stewart County were entirely the product of improvident southern land use traditions, and he was not so willing as others were at the time to see the chasms as expressions of a benighted South, a region where environmental degradation reinforced human degradation. As we will see, he was on to something.

Another regional progressive who grappled with Providence Canyon was Ellis Gibbs Arnall. Arnall was the governor of Georgia from 1943 to 1947 and one of the most intriguing figures in the state's twentieth-century political history. Arnall enjoyed a meteoric rise to political prominence in Georgia, becoming attorney general under Governor E. D. Rivers (who gave a speech at Providence Canyon in support of park designation) in the late 1930s, when he was only thirty-one, and then defeating Eugene Talmadge for the governorship in 1942, when he was only thirty-five. Arnall was one of the most progressive governors to rule the state during this period, but his open criticism of the state and region caused him trouble with his more conservative constituents. Among his many gubernatorial accomplishments, he repealed the poll tax, reformed the state's penal system, and cooperated when, in 1944, a federal district court judge ruled that the white primary, a key mechanism for disenfranchising black voters in the one-party South, was unconstitutional. For those and other apostasies, Eugene Talmadge and his allies attacked Arnall and regained the governorship in 1947 in one of the most controversial elections in the state's history.[53]

Along with his other progressive credentials, Arnall was openly critical of the land use patterns that had degraded much of Georgia. In a chapter titled "Land Laid Waste" in his book *The Shore Dimly Seen*, Arnall began with Providence Canyon. Admiring its "perverse beauty," he recalled a Sunday school lesson about how the earth was holy and ought to be respected as such. Providence Canyon was, to him, a major example of the failure to heed that moral imperative. "Where once there were fields of cotton and corn," he wrote, "was this great chasm. Within a generation, the unprotected land had been despoiled of its richness, then swept away, until there was a nothingness paneled in red and yellow and cream and a score of variations of these colors."[54] For Arnall, Providence Canyon stood for "the one crop system and its veritable mining of the soil." "It is not cotton that the Southeast is exporting to feed the hungry spindles of the world," he continued. "It is the very soil of the South. Year after year the fields are fed their starvation diet of guano; year after year the rains sweep the hillsides and wash away even greater wealth; year after year the yield of each acre of the tortured land is just a little less. Year after year the owner's profits grow smaller, and the hunger of the sharecroppers mounts."[55] Many residents of Georgia and the South resented such critical remarks, and Arnall's political career suffered as a result. But his indictment of Providence Canyon and its larger meaning was nonetheless a powerful one.[56]

Providence Canyon's most prominent appearance in the era's conservation literature came in Stuart Chase's 1936 classic, *Rich Land, Poor Land: A Study of Waste in the Natural Resources of America*. Chase was a prolific writer and popular economic theorist who contended with the social costs of waste and championed the need for planning as a way of tackling it. He was one of many radical interwar intellectuals enamored of planning for industrial and social efficiency, and insistent on the importance of technical expertise to rational planning. Chase focused mostly on industrial and labor issues during his early career, gaining fame for his investigations into the meatpacking industry in the late 1910s and early 1920s while he was a member of the Federal Trade Commission. After losing that job in a Red Scare purge of allegedly socialist employees, Chase was briefly involved in the Thorstein Veblen–inspired Technical Alliance after World War I, and he later founded the Labor Bureau, a research organization that marshaled facts in the fight for rational reform. It was while with the Labor Bureau that he began his sustained focus on waste. He enjoyed publishing success during the mid-1920s with *The Tragedy of Waste* in 1925 and the 1927 bestseller *Your Money's Worth*, which he coauthored with consumer advocate F. J. Schlink. Chase and Schlink went on to form Consumers' Research, the predecessor organization to the Consumers Union, the publisher of *Consumer Reports*. But it was in a series of books in the early 1930s—including *A New Deal* (1932), from which Franklin Roosevelt allegedly borrowed the moniker that would come to define his administration—that Chase made his strongest argument for a planned society controlled by philosopher engineers. Roosevelt became a great admirer of Chase and included him in his "Kitchen Cabinet," a group of informal policy advisors, and Chase became a popularizer of Roosevelt's New Deal programs and policies.[57]

During the mid-1930s, Chase turned his attention to natural resources and the social costs of their profligate use. His book *Rich Land, Poor Land* was the most prominent product of this interest. To illustrate natural resource waste, Chase chose as the book's feature image a glossy photograph of an erosion gully, which was placed opposite the title page as his frontispiece. The photograph, labeled "Plate I," was titled simply: "Gully in Stewart County, Georgia."[58] In a telling inversion of Bennett's juxtaposed images in *Soil Conservation*, Chase chose as "Plate II" an image of "Primeval Forest," again indicating that the gullies of Stewart County were the epitome of human-induced soil erosion in the United States, the antithesis of the continent clothed in its aboriginal verdure.

Later in *Rich Land, Poor Land*, and in an excerpt from the book published in *Harper's* under the title "Where the Crop Lands Go," Chase took his readers on an extended tour of Stewart County.[59] "When one becomes erosion conscious," Chase instructed, "a motor trip through the country, especially west of the Alleghenies and south of Washington, D.C., becomes an endless game of finding gullies. One spots them as a beggar spots a coin." "In pursuit of this grim game," he continued, "I once followed gullies to their supreme exhibit in this country—Stewart County, Georgia." As Chase and his party entered the county and passed through Lumpkin, he summoned every stereotype he could think of to signify southern rural poverty: they traveled "over a blood-red dirt road, past cotton fields awaiting the plow in March, groves of long-leafed pine, shacks that passed for farmhouses, bleak unpainted churches, jessamine vines, evil side gullies, a lovely square red-brick courthouse, mules, happy-go-lucky negroes, broken down Fords, deserted lumber mills, rickety cotton gins, crossroads stores whose whole stock seemed to be Coca-Cola and chewing tobacco." To the west of Lumpkin, the gullies got larger. "Suddenly, across a dipping valley," Chase observed, "we saw the hill on the other side laid open in a bright red scar 50 feet wide. 'That's one of 'em,' said our driver, 'but not the best one. You wait.'"[60]

After seeing a few smaller specimens, they arrived at Providence Canyon. "Presently the road approached a kind of isthmus, perhaps 100 yards wide," he wrote. "A plowed field was on the left and beyond it a sickening void. A battered church stood on the right, a few pines about it, then another void." The church he described was Providence Methodist Church. Then Chase's guide, a local CCC enrollee, went to work. "'Yes sir,' said our guide, 'Here's the old he one, the one on the left. He started the whole system. And do you know what started him? A trickle of water running off a farmer's barn about forty years

RICH LAND
POOR LAND

A Study of Waste in the Natural Resources of America

by

STUART CHASE

Maps, Diagrams and End Paper Design
by HENRY BILLINGS

New York WHITTLESEY HOUSE *London*
McGRAW-HILL BOOK COMPANY, INC.

ago. Just one damn little trickle, and now a third of the county's gone—forty thousand acres. Don't get too close to the edge. Sometimes she goes in an acre at a time.'" Chase asked what had become of the barn. The guide pointed to the bottom of the gully and told Chase that the barn as well as a schoolhouse and the church graveyard had slipped into its maw. "You see this road with a gully on either side?" he continued. "Well, they've moved it three or four times already, but this is the last time. Come summer and a few smart storms, and the field here will go, that church will go and the gullies will join and there'll be no more road. And how will you get to Alabama then?" The answer, of course, was that you would have to go around. "Yes sir, you will," his guide continued, "At least ten miles either side before you hit solid ground. There's nothing but nests of gullies north and south. This was the last way through."[61]

Chase was sickened, mostly, by what he saw as a profound example of wasted resources. Here was a supreme exhibit indeed. Yet as they came upon Providence Canyon, its beauty also impressed him. "Only once before have I seen a comparable phenomenon—the canyon of the Yellowstone in Wyoming. That was geological erosion, and even grander; this was manmade, but sufficiently superb." "The chasm," Chase admitted, "was awful and beautiful." He was charmed by the colors and hues of the canyon walls, and by the unusual pinnacles and other formations. "The earth strata changed from red to yellow to brown, mauve, lavender, jade, ocher, orange and chalk white. Pinnacles rose from the gully floor, sometimes with a solitary pine tree on their top at the level of the old land, banded and frescoed with color."[62] Chase's comparison of Providence Canyon to the canyon of the Yellowstone was certainly of a piece with the local boosters' view that it ought to be classed with the West's best erosion spectacles. Providence Canyon was sublime, Chase intimated, and its sublimity was what made it so confusing.

Whatever beauty Chase saw in the collapsing land was quickly counterbalanced by the surrounding landscape, and by his sense of the human processes that had created it. "The good earth had given up the struggle," Chase observed, "Yet here was a field plowed right to the edge." Chase wondered aloud whether the chasm's development might be stopped or even healed. His guide responded that it might be, if the runoff was diverted and those areas that could be were planted in kudzu and black locust, the era's favored botanical remedies for severe erosion. But it would be an expensive effort, he suggested. "What ought to be done with them?" Chase then asked about the giant gullies. "Well, sir," his guide responded, "I'd have the government buy up the whole county and turn it into a national park—with plenty of railings. You might even charge admission. You don't see a sight like this often." Chase responded to this suggestion with a muted expression of irony: "I looked into the vivid, slipping horror," he wrote: "'No,' I said, 'You don't.'"[63]

"Stewart County has gone," Chase concluded. "No work by man or nature can bring it back within a calculable future." He saw the gullied county not as a place to be saved but as a fate to be avoided. "Tens of millions of acres of American crop lands are taking the same precipitous path," he warned, "and no virgin west remains." There was the New Deal environmental critique in a nutshell. Then he invoked the beneficence of the New Deal conservation: "But all over the country groups of men like the conservation worker who drove us to Stewart County—lean, tanned men with clever hands and keen eyes—have set to work to check the landslide. Many farmers are aiding them. But they cannot do it alone. There must be more of them, and they must feel the force of public opinion behind them."[64] If anything was going to get the force of public opinion behind New Deal soil conservation programs, Chase hoped, it was the "supreme exhibit" of Stewart County's gullies, Providence Canyon foremost among them.

In this encounter between local guide and national reformer, the two interwar visions of what Providence Canyon meant collided. Like most of his cohort, Chase had made the canyon stand for a particularly southern variant of land abuse, though he was well aware that its analogues could be found in other regions. That Providence Canyon was a horrible monument to human carelessness and waste, and particularly to the political economy of southern agriculture, he did not doubt. Yet Chase conceded that there was something impressive there too, a beauty worth seeing. But the problem with making it into a national park, the irony even, was precisely in the distinction he made between geological and "manmade" erosion. National parks preserved nature, and particularly the supreme examples of wondrous natural processes: geologically eroded canyons, volcanoes, glacially carved valleys, waterfalls, unusual thermal features, and giant trees. More than that, they kept the works of humanity at bay, preserving a few remnant parts of the American landscape where natural processes would not be disrupted by human hands. Despite its eye appeal, then, Providence Canyon could not possibly pass as a national park because, as far as Chase was concerned, it was not natural.

If Stuart Chase provided the most substantial and entertaining introduction to the gullies of Stewart County to be found in the New Deal environmental literature, there is no better place to locate the denouement of Providence Canyon's interwar fame than in Robert J. Flaherty's documentary film *The Land* (1942). Flaherty invented the documentary genre of filmmaking with a series of ethnographic films in the 1920s and the 1930s. In fact, it was in a 1926 review of Flaherty's film *Moana*, which focused on a Samoan boy's coming of age, that the British filmmaker John Grierson coined the term "documentary."[65] Flaherty is probably best known for his groundbreaking 1922 film *Nanook of the North*, which followed an Inuit hunter and his family in their joyous struggle to survive along Canada's Hudson Bay. Flaherty also made several other well-regarded films, including *Man of Aran*, which profiled the lives of farmers on the Aran Islands off the coast of Ireland, and *Elephant Boy*, another coming-of-age story, about an Indian boy who matured into a brave hunter and elephant master. By the late 1930s, then, Flaherty was a renowned and successful documentarian.[66]

The Land was supposed to be the third in a trilogy of New Deal films about agricultural and resource conservation. Pare Lorentz, the head of the newly created U.S. Film Service, contacted Flaherty in 1939 to see if he would be interested in directing a documentary about the sorry state of American agriculture and about the various New Deal approaches for righting a ship in crisis. Lorentz had made a name for himself as a master of the documentary form with two Depression-era films of his own, *The Plow That Broke the Plains* (1936) and *The River* (1938), both of which he made under the auspices of the Resettlement Administration and its successor agency, the Farm Security Administration. *The Land* was to be another in that genre. But Lorentz was busy making a film about childbirth health and safety, *The Fight for Life*, and so he and the Film Service approached Flaherty, who, along with his wife Frances, accepted the assignment and moved back to the United States from Britain to get to work on the film.[67]

Russell Lord's *Behold This Land* was the inspiration for the film, and the Film Service soon hired Lord to help Flaherty write the script.[68] Robert and Francis Flaherty and their camera crew, which included Arthur Rothstein, then set out on three separate trips across America, traveling more than twenty thousand miles and shooting more than seventy-five thousand feet of film.[69] As Lord, who accompanied the entourage on some of their travels, recalled, the spirit of their first trip to Iowa and other points in the Midwest was "robust and cheerful." But as they moved into the cotton South, "the mood and temper of the party was not so happy." Flaherty was flabbergasted by "the condition of great stretches of cotton country." "It is unbelievable!" Lord remembers Flaherty lamenting, "unbelievable!"[70] One of the places they visited was Stewart County. In an early draft of the script, the gullies of Stewart County played a leading role. For a scene that was supposed to feature footage of "Gullies: Stewart County, Georgia," Lord's narration, in stanza form, read: "This was a beautiful farm once / on land of marvelous vigor. / Soil that it took the weather fifty centuries to make— / gone now / gone, in a century or less— / a clock tick in the span of eternity."[71] Had *The Land* followed Lord's script and walked in the footsteps of Lorentz's successful New Deal propaganda films, we might today know more about Stewart County's then-famous gullies. But circumstances intervened.

Copies of *The Land* are not easy to come by today, but if you can arrange a viewing you will see what appears to be Providence Canyon about a third of the way into the film, in a shot that pans from gully bottom up to gully edge. But the script had changed, and there was no mention of Stewart County in the film's final narration. It was one of a number of anonymous eroded landscapes that paraded by early in the film. The diminished role of Providence Canyon was likely the result of several formidable obstacles that the filmmakers faced in completing the project. Flaherty struggled to make a coherent film out of such a diffuse and overwhelming subject. Moreover, *The Land*, which had been influenced by John Steinbeck's novel *The Grapes of Wrath*, suffered in comparison to John Ford's stunning 1940 film adaptation. Indeed, *The Land* seems a documentary version of *The Grapes of Wrath*, but without the tight family focus that Steinbeck brought to his telling—and that had been a hallmark of many of Flaherty's other documentaries. Flaherty also failed to contain the tensions and contradictions of New Deal agricultural policy, which sought to provide relief to the most impoverished of the nation's farmers while simultaneously encouraging the mechanization and modernization of the agricultural sector in ways that pushed many off the land. Then, in 1940, a Congress increasingly hostile to the New Deal and building toward a war footing abolished the U.S. Film Service, leaving Flaherty to finish his film under USDA auspices.[72]

More than anything else, though, the nation's changing agricultural circumstances undid *The Land*. Flaherty had intended to make a film not only about "the waste of erosion and poverty," as he put it in his narration, but also about agricultural unemployment, the need for crop reductions, and what the New Deal was doing about those problems. But the war altered the complexion of American agriculture. The abused land was still there, but now there were strong impulses to increase production, and there were growing labor shortages in rural America as war industries drew workers into the nation's cities. With the help of his editor, Helen von Dongen, Flaherty tried to recalibrate the film to meet the new circumstances. But as one watches the film today, one can almost feel the foundations washing from under the whole enterprise. Some critics lauded the result. John Grierson allegedly called it the greatest film that Flaherty ever made. Most, however, have assessed *The Land* as flawed. To add a final insult to all of these injuries, the U.S. government pulled the plug on the distribution of the film after the Pearl Harbor attack, fearful that its critical history of American land use would provide ammunition for anti-American propaganda. *The Land* did enjoy a public premiere at New York City's Museum of Modern Art in 1942, but then the government buried it.[73] *The Land*, then, is less important as a film in which Providence Canyon made a prominent interwar appearance—it had but an anonymous cameo role, its potential star turn left on the cutting room floor. Rather, it is a parable about how and why Providence Canyon and the other rough, gullied lands of Stewart County receded from national consciousness after World War II. As the historian Finis Dunaway has written of *The Land*, "In its own way, the film marked the end of the New Deal."[74]

During the 1930s and early 1940s, Providence Canyon gave fame and focus to the fact of soil erosion. Soil erosion and the strong connections that reformers made between erosion and the socioeconomic conditions in the South were preeminent national issues during the interwar era, and they helped to make Providence Canyon America's best-known landscape of erosion outside of the Dust Bowl. Providence Canyon was everywhere in the era's soil conservation literature. Hugh Hammond Bennett led the charge, constantly invoking the gullies of Stewart County to successfully build a case for a national program for soil conservation. Many others followed Bennett's lead, making Providence Canyon and the rest of the county's gullies central to their soil jeremiads. Even as some, such as Stuart Chase, grudgingly admitted that there was a beauty to Providence Canyon, and as others, such as Jonathan Daniels, asked probing questions about just what had made the county's gullies so extreme, most focused on the simple fact that these huge gullies were dramatic artifacts of human land abuse in a region that was rife with evidence of such treatment.

After World War II, Providence Canyon's reputation as an environmental disaster faded. It still appeared in a smattering of books on soil science and soil conservation, and a number of environmental historians have made oblique reference to the place, but the invocation of Providence Canyon as a symbol of a seriously imbalanced regional agriculture became less coherent and compelling in the context of dramatic change.[75] The war shifted the economics of agriculture, ending the problems of overproduction and low crop prices that had spurred land planning and land use policies. The exodus of tenants, white and black, from the rural South and the coming of machines, chemicals, and the other trappings of modern agriculture severed the ties that had bound the region's social and environmental problems in the decades prior to the war. The congressional backlash against the New Deal and the Cold War culture of consensus dashed the reformist hopes of the era's conservationists, making their narrative of national sin and federal redemption less tenable. As New Deal conservationists found their agendas delimited within the United

States, many spilled out into the international development arena and helped to produce a New Deal for the developing world.[76] At the same time, federal agricultural policy increasingly favored well-to-do farmers while encouraging poor, inefficient producers to migrate from the land.[77] New synthetic inputs also changed the conversation about agriculture's environmental impacts and overshadowed the interwar era's concern for the nation's soils. "Permanent agriculture" became less relevant in the postwar years, and it was replaced by an "organic" ideal chiefly designed to counteract the environmental transgressions of the synthetic revolution. The Soil Conservation Service also shifted some of its troops to a new front: America's suburbs, where powerful new tools such as bulldozers and large-scale building techniques meant that soil erosion became a significant by-product of residential sprawl.[78]

Then there was the civil rights movement. For several centuries, African Americans in the South, most of whom worked the land, had experienced slavery and then tenancy partly as environmental predicaments. As a result, New Dealers put the human-land relationship at the center of their reformist vision for the region, and a few even took on the South's oppressive racial caste system in the process. But the postwar civil rights movement had, for good reasons, a stronger focus on legal equality and access to political power than it did on land reform. It was predominantly a movement of towns and cities, even as it attacked the violence, discrimination, and disenfranchisement in the Jim Crow South's most rural areas. Moreover, as Pete Daniel has argued, federal and state agricultural agencies and policies blatantly discriminated against African American farmers even as the civil rights movement was achieving its greatest victories. As a result, the African American farm population plummeted in the postwar years, undoing the connections between erosion and the African American experience.[79] Over the second half of the twentieth century, much of the former plantation South quietly reverted to forest, often with government encouragement and subsidies. Soon it was hard to see the gullies of Stewart County or their kin throughout the former plantation regions, for all the trees and other vegetation that scabbed over the open wounds of an earlier era.

By the late 1950s, even Providence Canyon's name had begun to fade from memory. In 1957, Edward Higbee, who taught agricultural land use and conservation at Clark University, published *The American Oasis: The Land and Its Uses*. Like many of his interwar predecessors, Higbee was concerned with soil and water conservation, but the American agriculture he described was a world away from the practices that had dominated in the plantation South during the nineteenth and early twentieth centuries. He mostly described an agriculture whose land base was contracting and shifting, whose land use was intensifying, whose farm population was plummeting, and whose productivity was increasing. He was still concerned about soil erosion, though mostly he hoped that technical solutions would mitigate the worst problems. Nonetheless, he occasionally invoked the nation's history of extreme soil erosion, using Providence Canyon, albeit under an unintended alias, as his supreme example. "Only rarely does soil erode in such an awesome manner as in Georgia's Paradise Canyon," Higbee wrote, getting the name wrong but referring to the place as if it was a common landmark. "But today," Higbee concluded of the southern farmer, "his new attitude is a reversal. Given time, many sections of the coastal plain promise to support some of the most prosperous soil-conserving agricultural systems in this country."[80] As this comment intimates, southern agriculture began to shift from the erosion-prone Piedmont and other hillier sections of the region to the flatter Coastal Plain, whose soils, though often sandy and lacking fertility, could be improved through fertilization and irrigation.

Edward Higbee's environmental concerns soon migrated away from American agriculture and into the nation's growing metropolitan areas, a move that paralleled the rural-to-urban migration of American environmental concern during the postwar era. In 1960, he published *The Squeeze: Cities without Space*, one of the era's most important critiques of suburban sprawl and the decline of open space, and he spent the next decade and a half writing about and advising decision makers on the urban and suburban land use issues that would powerfully inform the rise of modern environmentalism.[81] The New Deal had made a critique of American agricultural land use central to its conservation politics and policies. But the postwar environmental movement, focused as it was on issues of concern to an increasingly urban and suburban nation, largely averted its eyes from southern agriculture and its treatment of the soil. As it did, Providence Canyon lapsed from conservation consciousness.

As Providence Canyon faded into conservation obscurity during the postwar years, another group of meaning seekers, scientific creationists, appropriated it for their own purposes. Like the geological surveyors who first arrived at Providence Canyon in the early twentieth century, creationists have been interested in using Providence Canyon to support their own fantastical vision of the earth's history. What they see in the gullies of Stewart County is evidence of rapid historical erosion that supports their case that larger erosional landscapes, such as the Grand Canyon proper, could have been created in only several thousand years. As Rebecca Gibson insists in her article "Canyon Creation," which appeared in *Creation ex Nihilo* in 2000, "Providence Canyon beautifully illustrates how the geology of the earth is consistent with the short timescale of the Bible."[82] All of this suggests that while the local park boosters may have lost the battle over Providence Canyon's meaning during the interwar years to New Deal soil conservationists and their ilk, they could still win the war. Georgia's "Little Grand Canyon" may yet be the stuff of a national park.

PART THREE

Returning to Providence Canyon

THOSE WHO USED Providence Canyon as a visual symbol of extreme erosion built upon a deep tradition of seeing agriculture in the U.S. South as particularly destructive. Providence Canyon might still perform such a function today if visitors could look beyond their traditional expectations of a park as a place of nature left alone. Indeed, Providence Canyon seems a terrific place to define a new sort of park, one that honestly contends with the legacies, sometimes spectacular, of human land use on this continent. Why not make Providence Canyon the centerpiece of a park devoted to the centrality of soils and soil erosion to southern environmental history, a preserve that chronicles a crucial chapter in the history of American land *use* rather than one that celebrates how we have saved a few remnants of nature from such use? Instead of playing with the irony of Providence Canyon's status as a park, why not redefine what a park is in a way that defuses the irony? As such a park, Providence Canyon would have much to teach us about the intersections of nature and history in the South.

What lessons might we glean from such a park? From my first encounter with Providence Canyon, my sympathies have been with the New Deal conservationists and their sense of what these gullies meant. But there is also a contradiction at the heart of conservationist accounts and images of Providence Canyon that has nagged at me, a tension between its representativeness and its extremity. For Providence Canyon is at once a stunningly effective illustration of the human-induced soil erosion that plagued the region's agricultural lands and a freak of nature with few peers. To side with the soil conservationists and their view of Providence Canyon, then, one must make sense of why Providence Canyon became such an extreme spectacle. That, as we will see, makes for a more complicated story than either side was willing to tell.

Part 3 makes a case for how we might interpret Providence Canyon today. Chapter 6 focuses on Providence Canyon's representative attributes, delineating the sorts of broad lessons that New Dealers

{109}

hoped people would learn from the place, lessons that generally connected the South's agricultural practices and political economy with environmental destruction. But these lessons can only take us so far in understanding the erosive impacts of staple crop agriculture in the South. So in chapter 7 I turn to the environmental forces and qualities that contributed to widespread soil erosion. I also descend to the level of the local, examining several aspects of the gullying in Stewart County that fit less easily with regional generalizing and moralizing and that lend credence to some of the park promoters' claims about Providence Canyon's naturalness. Providence Canyon's interpretive possibilities are too rich and surprising to allow the place to languish any further.

CHAPTER SIX

Gullies and What They Mean

WHAT SHOULD WE SEE when we stare into Providence Canyon's multihued abyss? The first and most basic lesson that Providence Canyon ought to teach us is just how transformative southern agriculture was for the region's soils. While the Dust Bowl has garnered more historical attention as the signature soil catastrophe of frontier expansion in the United States, soil loss in the tobacco and cotton South was arguably a more substantial disaster. And while environmental historians have generally seen the Dust Bowl as the result of a mobile, mechanized, and highly capitalized form of frontier agriculture moving into an environment that, when disturbed, was susceptible to cyclical climatic conditions capable of lifting exposed topsoil by the ton into the atmosphere, the story of soil erosion in the South has been told as one of humans taking land ideally suited for farming and degrading it through a combination of primitive agricultural practices, inattention, avarice, and cruelty. By the 1930s, both regions—the southern plains and the tobacco and cotton South—saw streams of migrants leaving eroded land for towns, cities, and other regions, but the forces that propelled southern migrants off the land were different than those at work in the Dust Bowl.[1]

Let's first look the scope of southern soil erosion. Data from the 1934 Reconnaissance Erosion Survey provide a portrait of the nation's erosion problem at its peak moment of exposure.[2] The survey examined almost two billion acres and found that human-accelerated erosion marked more than half of that area. Almost 200 million acres, about 10 percent of the surveyed area, had been affected by serious sheet erosion, defined as the loss of more than 75 percent of its topsoil, and another 663 million acres, or 34.8 percent of the surveyed area, had suffered from moderate sheet erosion, or the loss of 25–75 percent of its topsoil. The survey also noted some evidence of gullying on 865 million acres of land, 45 percent of the surveyed area, severe gullying on 337 million acres, and more than 4 million acres of land destroyed by gullying. Much of Stewart County fell into this last category. Finally, the survey, which was completed just as the Dust Bowl was beginning, found 322 million acres adversely affected by wind erosion, almost 80,000 of which had been seriously damaged.[3] Speaking before Congress in March 1935 in favor of the Soil Conservation Act, Hugh Hammond Bennett usefully summarized the data generated by the Reconnaissance Erosion Survey. Excluding 360 million acres of the arid West, where "it was found impossible to differentiate accelerated or man-made erosion from normal geological erosion during the brief time available for the survey," Bennett estimated that 51,465,097 acres of land had been "essentially destroyed by wind and water erosion," and that another 105,594,229 acres had lost most of their topsoil. That's 157 million acres of American farmland rendered severely damaged by human-induced erosion.[4]

How much of this severely eroded land was in the South? Howard Odum, the distinguished sociologist from the University of North Carolina, crunched the survey's state-by-state numbers not long after they appeared, to produce a regional portrait in his important study *Southern Regions of the United States* (1936). Odum divided the South into two subregions, the Southwest (Texas, Oklahoma, Arizona, and New Mexico) and the Southeast (Virginia, West Virginia, Kentucky, Tennessee, North Carolina, South Carolina, Georgia, Alabama, Mississippi, Florida, Louisiana, and Arkansas). Of the 157 million acres of severely eroded land that Bennett had identified in his congressional testimony, Odum estimated that 97 million acres, or about 62 percent, were in the South, and of those 97 million eroded acres, probably 75 percent were in the Southeast. While erosion was by no means confined to the Southeast, the region clearly had more than its share of severely eroded land by the 1930s.[5]

The Piedmont saw some of the South's worst erosion. Several decades ago, the geographer Stanley Trimble, who studied the historical dimensions of soil erosion in the southern Piedmont, argued that during the height of row-crop agriculture, average soil loss was about seven inches *across the entire Piedmont*. This figure does not do justice to the spatial unevenness of Piedmont soil loss, but it is a stunning statistic nonetheless. In many parts of the Piedmont, Trimble argued, average surface erosion was closer to a foot in depth (see figure 13). In the early 1930s, Hugh Bennett had estimated that soil loss from Piedmont areas under cultivation had ranged from four to eighteen inches. Trimble's study supported the rough accuracy of those estimates.[6]

The distinctive red clay soils for which the Piedmont is known today are actually subsoils exposed as a result of the large-scale translocation of the region's topsoils. Bennett had grasped that fact as early as 1905 in his Louisa County, Virginia, soil survey. Returning to that point in a 1931 article, Bennett cited the example of Union County, South Carolina, just to the northwest of Fairfield County and today dominated by the Sumter National Forest, where soil surveyors mapped more than seventy thousand acres of "Cecil clay loam," a "new soil type" that was the "product of man-induced erosion." The case of Union County, Bennett suggested, was fairly typical of other Piedmont counties, where early soil types were often artifacts of transformative human activity.[7]

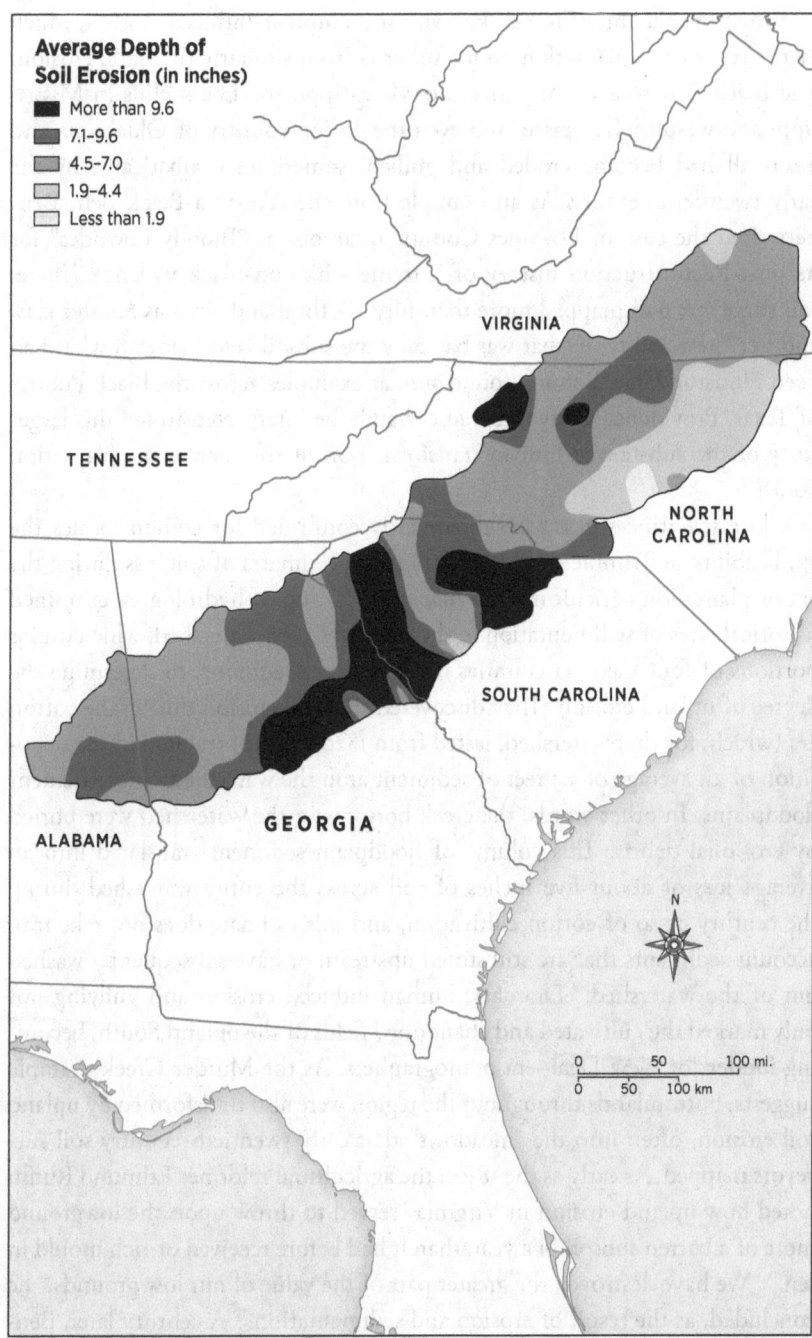

Figure 14. Average depth of soil erosion on the Piedmont. Adapted from Stanley Trimble, *Man-Induced Soil Erosion on the South Piedmont*.

While the Piedmont is well known for its human-induced erosion, much of the rest of the plantation South suffered from similarly dramatic erosion. The Black Belt soils of Alabama and Mississippi, the Loess Hills in Mississippi and western Tennessee, and even the cotton country of Oklahoma and Texas all had become eroded and gullied, sometimes dramatically, by the early twentieth century. As an example from the Alabama Black Belt, Bennett cited the case of Lowndes County, infamous as "Bloody Lowndes" for its post-Reconstruction history of extreme white-on-black violence. There, soil surveyors had mapped more than fifty-six thousand acres as Sumter clay, another "new soil type" that was basically the subsoil layer beneath what had been Houston Clay. Bennett found similar examples across the Black Prairies of Texas. Providence Canyon could certainly be interpreted to tell this larger story of the substantial human transformation of soils across the plantation South.[8]

Close scientific study has subsequently confirmed for certain locales the applicability of Trimble's and Bennett's broad estimates of soil loss during the era of plantation agriculture. For instance, a group of hydrologists examined historical rates of sedimentation in the Murder Creek watershed, which drains portions of four Georgia counties on the lower Piedmont, to determine the degree of upland erosion. They discovered that soil erosion during the cotton era (which, for that watershed, lasted from 1820 to 1930) resulted in the deposition of an average of 5.3 feet of sediment atop the watershed's presettlement floodplains. In other words, the creek bottoms in the watershed were buried by erosional debris. This volume of floodplain sediment translated into an average loss of about five inches of soil across the entire watershed during the century or so of cotton cultivation, and this estimate does not take into account sediments that are still stored upstream or have subsequently washed out of the watershed.[9] Dramatic human-induced erosion and gullying not only marked the cultivated and abandoned fields of the upland South, becoming fodder for New Deal–era photographers. As the Murder Creek example suggests, bottomlands throughout the region were also transformed by upland soil erosion, often into the "meadows" that early twentieth-century soil surveyors mapped. As early as the 1830s, the agricultural reformer Edmund Ruffin noted how upland erosion in Virginia "served to throw upon the lowground more of a barren subsoil in a year, than it had before received of rich mould in ten." "We have destroyed the greater part of the value of our low grounds," he concluded, as the result of erosion and sedimentation.[10] A century later, Bennett estimated that 50–60 percent of the Piedmont's bottomlands had been transformed by erosion sediment.[11]

The South's stream and river bottoms remain among the best places to witness and measure the long-term impacts of anthropogenic erosion. Most alluvial soils result from upland erosion and deposition, and they are generally rich as a result. But the transport of eroded soil and subsoil during the staple crop era was of a different order of magnitude. The erosion of upland sediment not only made streams and rivers turbid, but all that sediment raised streambeds and riverbeds substantially, particularly when it got caught behind small milldams. When that happened, several dynamics followed. First, water tables rose, making adjoining floodplains swampy. Moreover, because aggraded streams and rivers flooded easily, and watersheds stripped of their vegetation produced dramatic floods, overbank flooding often swamped and waterlogged surrounding lands. Finally, overbank floods deposited substantial amounts of sediment, often sand and gravel, on top of what had been fertile alluvial soils. During floods these sediments settled out in ways that augmented natural levees. Typically, as floodwaters escape their banks, they slow down and the sediments they carry settle out, with the heavier sediments settling closest to the banks. The result is the development of natural levees along watercourses. But during the plantation era, as sediment loads increased and aggraded rivers and streams flooded more often, floods produced enhanced levees, which trapped floodwaters in back swamps and prevented drainage back into streams and rivers. Such enhanced levees sometimes allowed stream channels to aggrade to the point that, within their augmented levees, they ran higher than the surrounding floodplains. The result was that alluvial lands became not only inundated with sediment but also permanently wet from a combination of higher water tables, more frequent flooding, and an impossible drainage situation. All of this sediment not only degraded the ecological conditions of streams and rivers, it also transformed once-fertile floodplains, lands that had been prized by both indigenous farmers and Euro-American settlers.[12]

Stafford C. Happ of the Soil Conservation Service pioneered the study of bottomland sedimentation in the 1930s and 1940s, and his work provides several examples of these phenomena in action. In the late 1930s, Happ conducted a reconnaissance of the bottomlands of the South Carolina Piedmont, and he found that 180,000 acres, or 60 percent of bottomland acreage in the region, had been made unsuitable for agriculture as a result of culturally accelerated sedimentation—what he called, echoing Charles Lyell, "modern" deposits—and the swamping that they had produced.[13] In another report, conducted with two coauthors, Happ studied the drainage basins of two creeks in Lafayette County, Mississippi, whose county seat, Oxford, is the home of the University of Mississippi. This study, meant to be a prototype for further studies, is significant because it takes us into a region known as the Loess Hills,

Figure 15. Walker Evans, "Erosion Near Oxford Mississippi," March 1936. This image captures erosion in the region of Stafford Happ's study. Courtesy of the Farm Security Administration / Office of War Information Photograph Collection, Library of Congress.

which run in a north–south belt through western Tennessee and Mississippi and border on the Mississippi Delta. Unlike the flat delta lands, the Loess Hills experienced substantial soil erosion and dramatic gullying in the late nineteenth and early twentieth centuries, iconic erosion that the photographer Walker Evans captured as part of his FSA duties (see figures 15–16). Erosion images from this region are among the most recognizable of the New Deal era.

Studying the twin creek basins between 1935 and 1937, Happ and his co-authors found extensive creek-bottom sedimentation as a result of upland agriculture, and the sedimentation had produced predictable results. The aggraded creeks raised water tables and increased overbank flooding, while sediment deposition on floodplain lands buried rich alluvial soils and trapped floodwaters. The authors did point out that the prominence of loess soils meant that at least some of the sediment was fertile. But the gullying that occurred proceeded through the loess and down into the sandy subsoils that

Figure 16. Walker Evans, "Erosion Near Jackson, Mississippi." This is one of Evans's most iconic images of soil erosion in Mississippi's Loess Hills. Courtesy of the Farm Security Administration / Office of War Information Photograph Collection, Library of Congress.

characterized the area, and the sediments debouched from the local gullies were the main culprit in covering alluvial soils with infertile sandy debris. Sedimentation from gullying, in other words, was much more damaging to bottomland soils than was sedimentation from sheet erosion, and it was a significant part of the total sedimentation that they measured. Happ and his co-authors estimated that about 40 percent of valley sediments had originated in gullies. Overall they found that 27.7 percent of the floodplain land in the two creek valleys had been damaged by swamping, sand overwash, or both, and that another 32.1 percent had become subject to frequent flooding and thus useless for agriculture.[14]

The 1940 Happ study of floodplain sedimentation and swamping made another intriguing claim, one that appears occasionally in the soil erosion literature of the time: that the swamping of bottomlands throughout the South as a result of upland erosion was exacerbating the region's malaria problems.

During the nineteenth century, Americans in most regions had suffered from malaria, but by the early twentieth century it was increasingly confined to the South. There a suite of factors combined to facilitate malaria's persistence: a favorable climate and ideal breeding conditions for the *Anopheles* mosquitoes that spread the disease, the proximity of large populations of rural laborers to these breeding areas, poor housing in areas where malaria was most prevalent, malnutrition and other chronic health problems that weakened human resistance, and even transformative human activities such as dam-building and reservoir creation. Observers often pointed to a strong correlation between cotton farming and the presence of malaria, the assumption being that poor tenant farmers and other cotton workers tended to live in swampy areas where the mosquito vectors bred with success. But Happ suggested that humans often contributed to those swampy conditions, providing mosquitoes with ideal breeding habitat. "Sedimentation," he concluded, "is an important factor in the cause or perpetuation of the swamp conditions that contribute to the problem of malaria control and eradication."[15] Arthur R. Hall, one of the most important early historians of soil erosion and conservation in the South, argued that settlers noticed this connection between bottomlands becoming "swampy and unfit for cultivation" and the growing incidence of "intermittent fever." Soil erosion, in other words, may have been making southerners, and particularly poor rural southerners, sick.[16]

While malaria retreated from the region after World War II, the other results of all of this sedimentation will be long lasting, even though upland erosion rates have declined precipitously since the 1930s as agriculture has contracted and forests have returned. By the early twenty-first century, for instance, the Murder Creek watershed in Georgia was more than 82 percent forested, much of it now part of the Oconee National Forest, and another 13 percent was in pasture, with only 2.3 percent in row crop agriculture. Erosive land use, in other words, had become minimal. Still, because floodplain storage of cotton-era sediments is thousands of times greater than the current annual rates of sediment export, it may take millennia for the creek to remove the accumulated sediment and return the watershed to conditions resembling those found prior to cotton agriculture, if it ever does. Until then, Murder Creek and its tributaries will continue to incise the accumulated sediment, its banks will continue to cave and erode, and the creek, like so many streams and rivers in the region, will remain muddy and sediment-choked long into the future. As the authors of the Murder Creek study conclude, "Generations of southeasterners have grown up under the assumption that Piedmont streams naturally featured steep unstable banks and turbid waters, while the reality is that these conditions are a direct long-term (multimillennial) consequence of poor farming practices."[17] A little more than a century of cotton culture, in other words, transformed the ecology, hydrology, and geomorphology of many southern watersheds in ways that may last for thousands of years.

One of the most remarkable places to encounter this history of soil erosion and floodplain sedimentation is Scull Shoals, an abandoned nineteenth-century mill town on the Oconee River in Georgia that also lies within the boundaries of the Oconee National Forest. I have come to think of Scull Shoals as a place with interpretive possibilities as rich as Providence Canyon's, for it too is a monument to the environmental legacies of the unsustainable cotton South. Settlers moving west founded the settlement that would become Scull Shoals in the 1780s. Many were Revolutionary War veterans given free land by the state in exchange for their service, and some were given additional land, under the headright system, for bringing slaves with them. In 1786, as a result of such settlement, the Georgia legislature created a new county along the Oconee and named it after General Nathanael Greene, who had commanded the Southern Department of the Continental Army during the American Revolution and did much to liberate the Georgia colony from the British. In 1793, in the wake an Indian attack that killed six settlers, the state fortified the Greene County settlement that would become Scull Shoals and called it Fort Clarke. This was the same year that Eli Whitney perfected his cotton gin, a revolutionary invention first tested on Mulberry Grove Plantation outside of Savannah, a property that had been given by the state of Georgia to none other than Nathanael Greene. The shoals, or rocky falls, that gave Scull Shoals its name thwarted upstream navigation, but they were attractive for mill development. By the end of the first decade of the nineteenth century there was already a gristmill and a cotton gin on the site, and soon there was a paper mill (likely the first in a state that would later become famous for its paper mills) as well as a toll bridge crossing the Oconee. In 1834, one of the first cotton mills in Georgia began operating at Scull Shoals, and by the late antebellum period there was a substantial brick manufacturing complex on the site, the Skull-shoals Manufacturing Company, that employed hundreds of workers and relied on a dam and race system that channeled fast-flowing water to animate two thousand spindles. By the eve of the Civil War, Scull Shoals was a vital industrial enclave into which cotton flowed from the surrounding region.[18]

Cotton was not the only thing flowing into Scull Shoals during the nineteenth century. While the community survived the Civil War and the ravages of Sherman's March (Sherman's troops burned the bridge across the Oconee and some baled cotton but left the mill complex standing), a quieter invasion was contributing to its demise. In the years before the war, as planters and their slaves cleared more of the watershed for cotton and corn production,

Figure 17. Scull Shoals ruins. These stone piers are the only visible remains of the Scull Shoals mill building and power plant. They were buried by anthropogenic sedimentation and now sit in a backswamp of the Oconee River. Photo by author.

sediment filled the Oconee's tributaries and then the Oconee River itself. As early as the 1840s, Greene County residents had begun to take notice of their eroding farmlands, and some began to organize agricultural societies to pursue reform. In 1851, Judge Garnett Andrews of Greene County warned area farmers and planters about the "fearful deterioration of our soil." "These bleak hills," he lamented, "with their gullied mouths, speak more eloquently and convincingly on the subject than I can." Sediment from Greene and adjoining counties headed downstream, building up behind the milldam and aggrading the river below the shoals to such an extent that it became difficult to get a sufficient head of water to power the mill's turbine.[19]

After the Civil War, as southerners remade the cotton plantation economy with nominally free labor, soil erosion increased in severity. As Arthur Raper, the New Deal–era rural sociologist who would make Greene County famous as the apotheosis of southern cotton tenantry, noted of this era, "The wasting soils from the fields continued to fill up the spring branches, pour into creeks and then into the Oconee and the Ogeechee. The fish were failing in the muddy waters. Floods and droughts were worse as forests were cleared and springs dried up."[20] Thomas Poullain, who owned the mill complex, and who had been one of the wealthiest planters in the state before the Civil War, struggled to keep the mill operating through the 1870s. By 1879, the mill property was auctioned off to pay his debts and back taxes, though the mill remained in operation under new ownership. Then came a disastrous flood in 1887 that swamped the entire community and undid what remained of the manufacturing operation. It was a flood that was almost certainly exacerbated by the morphological changes to the Oconee that came with erosion and sedimentation.[21] According to an 1889 report of the U.S. Army Corps of Engineers, the mill "failed from silting of the river below the dam."[22] The Scull Shoals community limped into the twentieth century, but its few remaining residents soon left.

Today Scull Shoals is a ruin, managed by the U.S. Forest Service as a historical and archaeological site. When one visits Scull Shoals today, several features immediately stand out. First, there is the curious fact that it is in the middle of the heavily wooded Oconee National Forest (I will return to that puzzle toward the end of the book). Second, if one walks away from the river, one easily finds remnant gullies now filled with old-field pine trees whose ages date the abandonment of cultivation. Most importantly, and thanks to the dogged removal of invasive Chinese privet that has overtaken so many narrow southern Piedmont bottomlands, one can still see the remains of nineteenth-century structures, including stone piers from the mill building, as well as the remnants of the bridge that once crossed the river. But there is absolutely no sign of the shoals, or of the dam that once rested atop their bedrock. In the mid-1990s, a landscape architecture professor at the University of Georgia, Bruce Ferguson, participated in a project, funded by the U.S. Forest Service, to figure out how to protect the historic Scull Shoals ruins from flooding, and he soon realized that the whole area had been inundated with sediment to a degree not previously appreciated. He came to suspect that the dam across the river might still be there, buried beneath sediment, and he hypothesized that much of the historical mill village was buried as well. Working with other University of Georgia scientists, who conducted subsurface magnetic and radar surveys for him (because the site is protected, they could not dig), Ferguson discovered strong evidence that his hunch was correct, including what appeared to be the old metal turbine several feet below what is today ground level. As a result, he posited that the visible remains of the mill building were in fact its second story, the entire first story having been engulfed by soil.[23]

With this evidence in hand, Ferguson reconstructed a history of sedimentation at the site. A water wheel powered the earliest mill, and it utilized the shoals, which must have had a drop of at least four feet, for its motive power. But by 1860, Ferguson speculated, sediment had aggraded the riverbed beneath the shoals and eliminated that drop. That year, the mill's owners built a new dam, which sat atop the bedrock of the shoals, that was ten feet tall, and water from the millpond that backed up behind the dam moved through races to power the mill. It was no longer the four-foot drop at the shoals, then, but the ten-foot drop produced by the dam that provided hydropower. By the time of the 1887 flood, the millpond behind the dam had silted up, perhaps to the top of the dam. The flood itself may have brought a debilitating pulse of sediment. Water continued to fall over the top of the dam into the twentieth century. Local residents recalled a falls of several feet at the end of the first decade of the century. But continuing erosion from early twentieth-century cotton production, and from abandoned cotton lands from the 1920s on, eventually aggraded the riverbed on the downstream side of the dam as well, obscuring the dam and further burying other parts of the site. Smaller dams downstream may have contributed to this dynamic. Today the river flows by the site sluggishly, with no sign of the dam. Ferguson estimated that sediment from the cotton era was at least fourteen feet deep on the downstream side of the buried dam and shoals, a figure that correlates well with what the subsurface surveys revealed of sediment depth at the mill complex.[24]

As in the Murder Creek watershed, one can expect that the Oconee and its tributaries will incise their way back down into this accumulated sediment, "making the streams their own polluters as they adjust back from the earlier devastation," as Ferguson has evocatively put it. One result may be the eventual reemergence of the dam and other parts of the mill site as well. But that may take a while, as tributary and upstream incision continues to deliver sediment to the site. "We should not expect a restoration, within any human time frame," Ferguson concluded, "of the Oconee to the conditions that Bartram witnessed two centuries ago." The geomorphological legacies of more than a century of extreme human-induced soil erosion will remain at Scull Shoals for a long time.[25]

What happened in the Murder Creek watershed and at Scull Shoals along the Oconee River happened throughout the region. Soil erosion produced heavy sedimentation that raised riverbeds, buried milldams, transformed river and stream ecosystems, and swamped rich alluvial lands with water and coarse sediment that rendered them useless for agriculture.[26] Tellingly, Stanley Trimble came to his study of Piedmont soil erosion only after an initial project on the geography of early water-powered industry in the South revealed scores of sites similar to Scull Shoals. As Trimble searched for old mills that he had identified through archival work, he became confused "when many sites turned out to be swampy morasses rather than the dams and associated installations I had expected."[27] It was at that point that he realized that huge volumes of erosion sediment had buried many of his proposed research sites.

The extent of these transformations came to light in the early twentieth century not only as a result of soil surveyors incorporating "meadow" lands into their mapping and soil scientists beginning to study the problem, but also as southern states passed laws encouraging the creation of drainage districts. The case of Georgia is instructive. In 1911, the Geological Survey of Georgia published a bulletin that surveyed conditions in the state and urged the passage of state legislation to create drainage districts. The state's wetlands below the fall line, and particularly the Okefenokee Swamp and the state's expansive coastal marshes, provided the main justification. The report hardly mentioned other parts of the state.[28] This was a time when such wetlands were not valued as they are today. Not long after the publication of the 1911 report, the state passed a law that allowed farmers to band together to create drainage districts and tax landowners within them to pay for drainage improvements. There was great hope that such a plan would "improve" some of the state's most intractable wetlands. But a funny thing happened after the law passed: overwhelmingly, the districts that the state's farmers organized were in the Piedmont. In 1917, when a follow-up report appeared, it noted that thirty-seven of the first forty drainage districts set up in the state were Piedmont districts. "The most serious drainage problem presented in the Piedmont section is the filling of the stream channels by the products of the erosion of the cultivated slopes," the authors of a 1917 report wrote. "This has in numerous cases progressed to a point where the adjoining rich bottomland is rendered wholly uncultivable."[29] The bottomlands that early settlers had considered the best lands for cultivation were, by the early twentieth century, prime candidates for drainage reclamation.

Providence Canyon could be made a vehicle for recounting this larger history of soil erosion in the upland South and the dramatic consequences for the region's bottomlands. Bruce Ferguson, who studied the Scull Shoals site, attested to this when he sought an example for the dramatic erosion that had layered so much debris atop the region's riparian corridors. "When runoff concentrated, it dug gullies," Ferguson instructed. "At its most catastrophic, runoff dug what is now called Providence Canyon near Columbus, Georgia."[30]

Who caused all of this erosion, and what are the historical lessons to be gleaned from an understanding of these causes? If we are to agree with the visitor center interpretation that Providence Canyon is "a spectacular testimony to man and his mistakes," what exactly were those mistakes, not only in Stewart County but also across the entire plantation South?

To begin with, the environmental destructiveness of southern agriculture was the product of a settler society pushing relentlessly westward and clearing land. Theirs was not the first agricultural society to have had an impact upon the regional environment. In few parts of North America north of Mexico did native groups more actively transform the land through their agricultural practices than did the indigenous peoples of the Southeast. This was no pristine wilderness into which settlers moved. But settler agriculture, as the cutting edge of an emerging global capitalist economy, was driven by a land hunger and market orientation that Indian agriculture never expressed. Not all staple crop producers in the South participated in this process of rapid frontier expansion into the region's upland forests and prairies. Specialty crops such as rice or Sea Island cotton could only be grown profitably in narrow environmental niches.[31] But tobacco and, later, short-staple cotton (which, like tobacco, was often accompanied by corn) were both commercial crops feeding transatlantic markets, and they tore through fresh land with a pace and magnitude that native agriculture never matched. This process began in the seventeenth century, as tobacco spread throughout the Chesapeake tidewater; it continued into the mid-to-late eighteenth century, as tobacco planting moved up onto the Maryland and Virginia Piedmont; and it culminated in the nineteenth century as cotton culture expanded to the south and then west. At every step, though to varying degrees, accelerated soil erosion corresponded with—and was the product of—expanding settler agriculture. The forces propelling this expansion ought to be at the center of Providence Canyon's interpretation.

Soil erosion in the South began with the removal by agricultural settlers of the presettlement "vegetative blanket" that had anchored soils in place.[32] Once exposed to the erosive effects of rainfall, these soils began their migrations downhill and into the region's streams and rivers.[33] This correlation between land clearance and destabilization of soils was true of most settler societies, and as we will see, land clearance alone had powerful impacts upon the region's hydrological cycle. The severity of erosion in the Southeast, however, also had to do with the specific ways that migrants first settled the region and which lands they tended to clear. Settlers brought a preference for rich bottomlands, particularly those covered in cane, and an aversion to the well-drained pinelands whose soils they assumed would be too droughty and nutrient poor for profitable farming. In the middle of that spectrum of assumed fertility were lands clothed in deciduous forest. These preferences meant that agricultural settlement was uneven throughout the region, with planters and farmers generally avoiding the flatwoods of the Coastal Plain for the more rolling portions of the South such as the Piedmont. These lands were covered in the sorts of mesophytic, or water-loving, vegetation such as oak-hickory forests that settlers saw as optimal. In fact, after the American Revolution, when the state of Georgia adopted a system of land classification for tax purposes, state officials based it partly upon the types of native vegetative cover. After bottomlands, "oak-hickory" lands were the most valuable category, while the tax code designated pinelands as the least valuable. As a rule, settlers focused on clearing and farming undulating oak-hickory lands rather than flatter pinelands. "This meant," as Arthur R. Hall wrote, "that in many cases the first land to be taken up was the most rolling and the most exposed to erosion."[34]

Another factor critical to understanding pronounced soil erosion in the plantation South was the dominance of row crops such as cotton, tobacco, and corn. These crops were grown in widely spaced rows, often hilled or mounded, with the ground in between the rows aggressively tilled throughout the growing season to eliminate weeds and maximize crop productivity. The result was a lot of disturbed and exposed soil prone to erosion.[35] Many also blamed the crops themselves, particularly tobacco and cotton, for being demanding of soil nutrients, though a number of historians have also pointed out that corn may have been the worst offender when it came to producing erosion.[36] Others connected the region's erosion problems to poor agricultural technique. As numerous commentators attested, particularly during the early national period when plow agriculture became widespread, southern farmers and planters often plowed and planted up and down slopes rather than across them, which concentrated runoff in downhill furrows (and sometimes, in corn culture, they cross-plowed, effectively doing both). Agricultural reformers bemoaned the practices of these "down hill farmers," who they blamed for the substantial erosion and gullying that was increasingly obvious in the region's fields. Certainly some of the region's history of soil erosion can be blamed on such approaches to farming on rolling land and to the neglect of soil conservation measures that might have mitigated such damage.[37]

Such an indictment, however, misses the extent to which some southern farmers and planters *did* develop and use conservation techniques to control erosion. In the early years of American independence, "horizontal plowing" (what we today would call contour plowing) spread from Europe into Virginia and then slowly south into the cotton states. Thomas Mann Randolph, Thomas Jefferson's son-in-law, is usually credited with importing the practice in the 1790s and early 1800s and with inventing a hillside plow that facilitated it, and Jefferson himself became a strong advocate of horizontal plowing. By the 1830s, horizontal plowing had spread, albeit unevenly, from Virginia down through the Carolinas and into Georgia. Horizontal culture was also widely practiced in the Natchez District and the central parts of Mississippi, a region settled ahead of the general westward move of cotton as a result of the access provided by the Mississippi River and market opportunities in nearby New Orleans. Indeed, it was Jefferson who encouraged a prominent Natchez District planter named William Dunbar to adopt horizontal plowing, and Dunbar spread the practice throughout the state.[38]

Horizontal culture, although avidly practiced in certain areas, was not widely adopted across the region during the antebellum period, and not only because of ignorance, inattention, or hostility among planters and farmers to the ideas of elite "book farmers." Rather, horizontal plowing was not the panacea that reformers hoped it would be. For some the problem was poor implementation, or an unwillingness to experiment fully enough to adapt the technique to a particular piece of land. Others lacked the necessary technology. But there was a more fundamental problem: while contour plowing worked well on gentle slopes and under conditions of light to moderate rainfall, heavy rains falling on steeper slopes stressed contour furrows and caused breaches that carried with them substantial amounts of soil. While horizontal plowing could control the gravitational energies of rainfall to a point, once those energies reached a high level of intensity, horizontal furrows became a liability from a soil conservation perspective.[39] As Arthur Hall concluded, "Horizontal plowing was criticized because the capacity of the water furrows was often not great enough to hold the water of a heavy rain. When this circumstance did occur and the ridges broke, the accumulated water caused accelerated washing."[40] Indeed, when heavy rains drained in a concentrated way through breaches in contour furrows, gullies ensued.

One result of the failure of contour plowing under such conditions was that a number of reformers began advocating the placement at regular intervals of hillside ditches, which could drain excess water from the slopes of fields and relieve pressure on contour furrows. Ditching was slow to catch on in Virginia and other tobacco regions, but it became relatively popular in the cotton states in the decades before the Civil War. Richard Shipp Hardwick of Hancock County was Georgia's great champion of ditching. But ditches needed to be properly engineered or they too could become foci for washing and gullying. As Hall put it, "unless some precaution was taken with the outlet, the process of gullying might be accelerated rather than prevented by hillside ditches." Agricultural historian John Hebron Moore concurred: "Although ditches permitted rain water to run off without overflowing the tops of the rows, they still did not solve the problem of preventing gullies. Water drained off the fields and rushed down the hillside ditches in large torrents and in time the ditches themselves turned into great gullies." Farming on the contour in the nineteenth-century South was technically challenging, and it often created soil washing that was as bad as, if not worse than, "down hill" farming. Under the circumstances, some of those who worked the hilly lands of the region may have concluded that downhill farming was the more salutary proposition from a drainage perspective. "With such confusion and lack of mastery of the technique," the historian of Georgia agriculture James Bonner noted, "it is little wonder that many Georgians adopted the system slowly and with misgivings."[41]

Hillside ditching was the precursor to a system of terracing that began appearing in the 1840s and took hold over parts of the region in the late nineteenth and early twentieth centuries. Terracing involved creating wide strips of land on the contour, often with a drainage ditch incorporated into the terrace. Terracing spread throughout the Piedmont between 1880 and 1920, a development partly fomented by the growing number of state and federal agricultural experts who came with the land grant colleges, the experiment stations, and eventually the extension service.[42] Recall Hugh Bennett's memories of helping his father build terraces on their Anson County, North Carolina, farm.[43] Today one can still find terrace remnants throughout the undulating portions of the plantation South, though now they are usually covered by woods or appear as faint contour lines on pasture lands. But, again, poorly constructed and improperly maintained terraces also could become a primary cause of accelerated erosion rather than its putative solution. The geographer Carl Sauer argued in the 1930s that the "present erosion crisis in the Southeast is the result primarily of the introduction of terracing, originally thought of as protection against erosion." Sauer likely exaggerated the blame that terracing deserved, but his observation nonetheless suggests that terraces were causing erosion problems, and particularly gullies, as late as the early twentieth century. The authors of a 1930s study of gullying in Spartanburg County, South Carolina, concurred. "Many of the deep gullies of the Spartanburg area," they concluded, "started from terrace drains."[44] Beginning in the 1930s, the Soil Conservation Service improved terracing technique, but by that point the plantation South was in decline and its farmers were moving away from the region's most sloping lands.[45]

While it might be tempting to see a spectacular gully like Providence Canyon as a symbol of a regional neglect of soil-conserving practices, then, such an interpretation would miss the significant history of soil conservation in the region. Certainly many were neglectful, and the consequences for the region's soils were serious. But Providence Canyon might just as well represent a subtler lesson: soil erosion occurred in the plantation South not only because some farmers and planters rejected soil-conserving practices but also because the efforts that many did make to conserve the soil, efforts that were substantial, and even pioneering, could do as much harm as they did good. Spectacular gullying was one of the signature results of sincere but less-than-adequate attempts by farmers and planters to channel runoff, terrace and drain their fields, and protect their soils.

Another frequent technical complaint was that southern farmers and planters worked with inferior tools and turned the soil too shallowly. This line of lamentation began with the southern plantation's heavy reliance on the hoe as its chief agricultural implement, for the hoe came to have strong negative connotations among improving planters and outside observers alike. The central place of hoes and hoe-like implements in American Indian and African agriculture led many to make developmental distinctions between primitive hoe cultures and advanced plow cultures.[46] Even where plows were used, they were often used badly. James Madison combined two prominent critiques when, in the early nineteenth century, he wrote: "Shallow plowing and plowing up and down the hilly lands, have, by exposing the loosened soils to be carried off by rains, hastened more than anything else the waste of its fertility."[47] Madison was by no means alone in this indictment.

The main type of plow used in the late eighteenth-century and early nineteenth-century South was a light shovel plow, usually led by a single horse or mule, which only turned the soil to a depth of three to four inches. Such one-horse plows, like hoes, were well adapted to conditions on newly cleared land, which often still had dead trees, roots, stumps, and other debris to negotiate. (As we will see, a tremendous amount of southern farming was done on newly cleared land). Nonetheless, observers blamed shallow plowing for erosion because it did not permit subsurface infiltration and increased surface runoff. Reformers thus constantly advocated deeper plowing so as to break up the subsoil and allow greater rainfall infiltration and moisture retention. Many also thought deeper plowing could counter nutrient depletion by bringing new sources of fertility into the crop zone. But when, after the American Revolution, farmers did begin to adopt heavier moldboard plows with iron shares, which usually needed to be pulled by multiple draft animals, many complained that the relatively infertile subsoils of the region became mixed with the topsoils in ways that made them less productive. Even though shallow plowing could be a major culprit in accelerating surface soil erosion, deep plowing was not always an attractive solution. The development of the substratum or subsoil plow, which could break up the subsoil without mixing it with the topsoil, solved some of these problems and became an important technological adaptation beginning in the 1820s. But for several reasons, their expense and the additional stock they required preeminent among them, such tools and techniques were only sparsely used. The technological primitiveness of southern agriculture was thus a constant source of comment well into the twentieth century, and it may have been a culprit in accelerating erosion in the region. But as we have already seen, improved plowing techniques could also be highly erosive.[48]

Southern agriculture was also extensive and itinerant. Rather than sticking in one place and maintaining, or even building, soil fertility over time, southern farmers and planters tended to work soils until their fertility had dissipated and then move on to fresh lands. Instead of practicing continuous cultivation on farms with permanent fields, what Franklin King and Cyril Hopkins would later call "permanent agriculture," a significant percentage of southerners, from the largest planters down to yeomen farmers on the margins of the plantation system, practiced shifting cultivation. As Lewis Gray, the agricultural historian and New Deal reformer, memorably put it, "The planting economy was based on deliberate exhaustion of the soil and the expectation of making from one to three moves in a single generation. Planters bought land as they might buy a wagon—with the expectation of wearing it out."[49]

Shifting cultivation was not uncommon in other regions of North America where frontier conditions prevailed. Where land was cheap and labor dear, there were strong economic incentives for approaching native soil fertility as an account to be drawn down. The historian Avery Odelle Craven made this point explicitly in his formative study, *Soil Exhaustion as a Factor in the Agricultural History of Virginia and Maryland, 1606–1860*, which was first published in 1926. Craven, who had completed a master's degree under the supervision of frontier historian Frederick Jackson Turner before heading to the University of Chicago to complete his PhD, rejected arguments that soil exhaustion was distinctive to the South and instead maintained that what happened in Virginia and Maryland, the focus of his study, was of a piece with settler agriculture in other frontier contexts. "Frontier communities," Craven insisted, "are, by their very nature, notorious exhausters of their soils." In fact, plaints about "worn out soils" could be heard throughout the nation during the early nineteenth century. But despite Craven's protestations to the contrary, southerners did practice shifting cultivation to a larger extent and for a longer time than did farmers in other regions. If shifting cultivation characterized frontier agriculture, then the frontier lingered in portions of the South in troubling ways.[50]

More than any other southern agricultural habit, shifting cultivation, what many critics came to call "land skimming" or "soil mining," has anchored environmental critiques of southern staple crop production, although how much blame it deserves for causing soil erosion remains a thorny question. Many have argued that southern farmers and planters were eager westward migrants quick to search out fresh lands when theirs ceased producing and who left in their wake wide swathes of abandoned and eroded land growing up in broom sedge and old-field pines. Some have suggested that these bad habits began in the early colonial period, when primitive agricultural techniques seemed expeditious, and settled into an ingrained traditionalism that was hard to get southerners to give up. Others have cited a land-labor ratio that made shifting cultivation economically attractive in an almost deterministic way. Whatever the explanation, the moral of the story for many early observers and some later scholars has been that this migratory approach to managing soil fertility was negligent, primitive, and unsustainable.

Land policy in the region certainly encouraged extensive agricultural practices. Georgia provides a stellar example, for as historian Paul Wallace Gates put it, "the rapid expansion of cotton planting in Georgia was facilitated by the prodigal distribution of public lands."[51] At the end of the American Revolution, Georgia had a public domain that stretched all the way to the Mississippi River, and this vast stretch of land became the subject of tremendous speculation, including, most famously, the Yazoo Land Fraud of the 1790s, when Georgia sold off thirty-five million acres of its western lands in a transaction riddled with corruption and bribery. In 1802, the state ceded its territory west of the Chattahoochee River to the federal government. Then, fatefully, it created a land lottery system for disposing of the rest of the public domain within its newly drawn boundaries. Over the next several decades, Georgia sponsored eight land lotteries as settlers pushed back the Creeks and Cherokees, with federal assistance, and opened their lands to colonization. Many contemporaries complained that the inexpensive lottery lands encouraged southern planters and farmers to practice shifting cultivation and emigrate when the lands they were farming played out.[52]

As early as the 1830s, central Piedmont lands, which had only just been opened to settlement during the three previous decades, showed signs of serious wear. In 1834, a farmer in Putnam County, Georgia, noted, "Already throughout the state are wasted plantations, fields turned out and grown up in broomsedge." Blaming the land lottery, Solon Robinson, the well-known agricultural journalist and reformer, wrote of the state's lands, "Probably no soil in the world has ever produced more wealth in so short a time, nor been more rapidly wasted of its native fertility." A Sparta, Georgia, farmer wrote that the lottery "begat a careless, slovenly, *skimming* habit of farming." In 1851, John Trott of Troup County, north of Columbus, spoke of the "waving broomsedge, the barren hillsides, and the terrible big gullies" that had come to mark the agricultural landscape of his region. By 1859, along the lower Chattahoochee Valley not far from Stewart County, another local observer noted evocatively that the land had begun to "put on the wrinkles and furrows of premature old age brought on by a soil-killing culture."[53]

But not everyone has agreed with this glum assessment of shifting cultivation. Despite frequent charges that it left exhausted and eroded soils in its wake, a number of historians and geographers have argued that shifting cultivation could amount to a systematic and sustainable approach to managing soil fertility. These scholars point to sophisticated forest fallow approaches to farming that were difficult for the uninitiated to appreciate. Under such systems, farmers and planters cleared fresh land, often with the help of fire, which converted the surface vegetation to ash and provided soils with a pulse of nutrients. They then planted the newly cleared land with crops, usually tobacco or cotton, sometimes preceded or followed by corn or other grains, for several years until the land's fertility gave out. At that point, they would leave that plot to the processes of "old field" succession and clear another one, beginning the cycle anew. At first glance, this looked a lot like soil mining, but there was one key distinction: under these systems, farmers eventually circled back to these fields after a long fallow period of up to twenty years, which allowed for a new growth of forest that helped to restore the soil. The results were not pretty, at least to those whose aesthetic was one of a densely settled agricultural countryside marked by permanently cleared and improved fields. Shifting cultivation looked like Indian agriculture, and it was often practiced with the hoes and other basic agricultural implements that prevailed in the region. Nonetheless, it may have worked reasonably well as a method for managing soil fertility within a given property or area. The question remains, however, whether such an approach was actually sustainable and, if so, for how long. Even those southern farmers and planters who practiced the most virtuous variants of forest fallow cultivation rarely celebrated them as ideal strategies in the ways that more recent scholars have. Perhaps this was because shifting cultivation went against their strong cultural preferences for a settled and permanent existence. But many also clearly recognized that shifting cultivation was less than ideal from an economic standpoint and that it seemed to be leaving damaged land in its wake.[54]

Employing shifting cultivation as a strategy for maintaining soil fertility required large land reserves. Planters, if they owned sufficient land, could engage in shifting cultivation within the bounds of their properties. The historian Steven Stoll cites the example of Thomas Cassells Law, who owned 5,500 acres in the Sand Hills near Darlington, South Carolina, but who averaged only about 400 acres in cultivation during the 1840s and moved those 400 acres through his property. At such a ratio of cultivated to unimproved and fallow land, Law might have been able to practice shifting cultivation indefinitely within the bounds of his plantation. Stoll evocatively describes it as "emigrating while staying in place."[55] Smallholders often tried the same approach, though under much greater stress from growing family size and limited land resources. Indeed, most small producers lacked the land base to emigrate in place for very long and thus had to move to new areas with fresh land more frequently. They also usually found themselves working poorer soils than the large planters, and like planters, they were often tempted by the echoing promise of untapped lodes of soil fertility farther west. Here the mining metaphor worked well not only to explain the results of shifting cultivation but also to evoke the shimmering promise that propelled it.

This persistence of shifting cultivation was a distinctively southern pattern. The economic historian John Majewski has noted that agricultural census data on improved and unimproved farmland, which the Census Bureau began collecting in the mid-nineteenth century, showed that the percentage of unimproved land within farms was consistently higher in the South than it was in other regions. In 1860, for instance, farmers in the Northeast and Midwest improved almost two-thirds of their farmland, while southerners improved only about one-third of theirs, statistics that Majewski took to be a clear indication of the persistence of shifting cultivation in the South. While he concluded that shifting cultivation was a sensible strategy for managing soil fertility, he also saw its persistence as a major reason for the region's economic underdevelopment. Sparse rural populations, large numbers of slaves who consumed few marketable resources, and the small percentage of land in production at any one time created economic "dead zones" that made it difficult for the region to support urbanization, modern transportation networks, industrial production, and thriving markets.[56]

During the antebellum era, outside observers constantly entreated southern farmers and planters to give up their itinerant ways, settle in place, and achieve a landed permanence. This group rarely appreciated that shifting cultivation could be a strategy for managing soil fertility. Among the most famous and influential of these observers was Frederick Law Olmsted, the landscape architect of Central Park fame, who, in his travels through the South, heartily disapproved of the ubiquitous signs of the southern attachment to shifting cultivation.[57] But the more vocal and incisive critics were not northern and European outsiders but southern agricultural reformers who worried about the myriad costs of constant emigration and land abandonment. For some the problem was fundamentally political. At a moment when northern and southern states battled over critical issues that would move the nation toward civil war, some southerners worried that the sparsely populated agricultural landscapes required by shifting cultivation and the constant loss of population to westward migration would diminish their political power. Shifting cultivation was a political problem, then, which helps explain why many of the region's most important agricultural reformers were also prominent defenders of slavery and guardians of regional political power. For others, settling in place and building a landed permanence was a cultural imperative, and the failure of the region's farmers and planters to do so seemed a profound regional character flaw. For those whose ideal of agricultural improvement involved visible environmental mastery, shifting cultivation looked feral and uncivilized, and it raised concerns that southern society was degenerating toward the primitive. For those who saw shifting cultivation as a marker of frontier conditions, the region seemed developmentally stuck. While southern agricultural reformers thus pleaded with their fellow farmers and planters to practice sedentary soil stewardship, the permanence they thought such stewardship would provide was a means to a larger set of political, economic, and cultural ends, not an environmental end in itself.[58]

Most antebellum agricultural reformers argued that southerners, in order to shake their shifting habits, needed to adopt mixed husbandry, the critical component of which was the return of organic nutrients to their soils. In a system of mixed husbandry, farmers combined the growing of crops with intensive livestock husbandry, and while livestock could provide numerous resources and services—meat, dairy, leather, and motive power among them—the most critical was dung, or livestock manure. The careful production of dung and its application to cultivated fields returned nutrients to the soil that crop production had removed, and as an ancillary benefit, its organic content lent soils greater structural cohesion that reduced the potential for erosion. But to produce sufficient dung to keep fields fertile indefinitely, farmers and planters had to dedicate themselves to raising healthy livestock, which in turn limited the acreage that they could plant in staple crops. They had to improve pasturelands so that stock were well nourished and could produce high-quality manure, and in some systems they would need to rotate fields between pasture grasses and crops. In other words, they had to devote a substantial acreage to reproducing soil fertility rather than producing cash crops, and they had to devote substantial labor to tending to livestock, to collecting and applying manure to their croplands, and to maintaining a diversified farmscape. The maintenance of soil fertility on a single farm or plantation under a system of mixed husbandry required that farmers forsake profit maximization and staple crop specialization for long-term diversified production that necessitated more complex and varied labor regimes. For a variety of reasons, this was not an approach that many southern producers of staple crops chose to take, particularly as cheap western lands beckoned. As Thomas Jefferson succinctly put it in a letter to the British agricultural writer and agrarian reformer Arthur Young, "We can buy an acre of new land cheaper than we can manure an acre of old."[59] In the antebellum South, that economic logic usually reigned supreme.

As southern agricultural reformers argued for the adoption of crop-livestock mixes and the use of manure to sustain soil fertility, they had to contend with a unique southern institution that seemed both of a piece with the extensiveness of the region's farming and anathema to the goals of mixed husbandry: the open range. Beginning with the earliest settlement of the Chesapeake colonies in the seventeenth century and extending throughout the region during the eighteenth and nineteenth centuries, southerners largely let their livestock fend for themselves on lands beyond their cultivated fields. In the South, a tradition of commons grazing on unimproved land developed that allowed even smallholders or landless herders to raise cattle and hogs by grazing them on lands they did not own. Stock thus roamed an open range, so named because of legal regimes that required farmers and planters to fence in their crops rather than their livestock. If farmers and planters wanted to keep animals out of their fields, it was their responsibility to do so, not the owners of the stock. Like shifting cultivation, the open range seemed an adaptation not only to a region of abundant land and scarce labor but also to a landscape in which a substantial portion of the agricultural land base was out of crop production at any given moment—abandoned, fallow, yet to be cleared, or deemed too poor to produce crops in the first place. Many have seen open range practices as the product of a regional obsession with producing profitable crops such as tobacco and cotton at the expense of careful livestock husbandry. And to those who lamented the poor condition of southern soils, the open range has seemed an institution willfully wasteful of manure.[60]

The challenges that the open range posed for southern reformers interested in promoting mixed husbandry were obvious. Most importantly, it was almost impossible to collect the manure from free-roaming animals and apply it to croplands. Moreover, the quality of the forage on the southern open range was vastly inferior to the forage to be found on the improved pastures that the proponents of mixed husbandry encouraged. As a result, the region's livestock were, as a rule, small, lean, and marked by prominent physiological changes that resulted from their feral lives.[61] The manure that open range livestock produced, even had it been collected, was poor in quality and insufficient to the task of keeping fields fertile for long. Southerners also seemed uninterested in improving their livestock through selective breeding, or of sowing their pastures with imported grasses and fodder crops. For observers and reformers who were used to seeing careful livestock husbandry as a symbol of virtuous agricultural improvement and a civilized mastery of nature, the sorry stock on the South's open range were a sad testament to the unwillingness of the region's planters and farmers to improve their practices. The same was true of the general absence of barns and other infrastructure of a settled agriculture based on careful livestock husbandry. When Jonathan Daniels, in his confusion on the edge of Providence Canyon, invoked a "barn smelling of animal husbandry," he was tapping into a deep rhetorical tradition. Even the story of a dripping barn roof precipitating Providence Canyon thus had a redemptive quality to it, for barns were the consummate symbols of mixed husbandry. No farmer with a barn would have willfully ruined his soil. Nonetheless, in much of the region, the open range lasted into the late nineteenth or early twentieth centuries as an important partner to shifting cultivation. As a result, reformers' efforts to get southern farmers to settle in place and improve their soils also foundered on southerners' unwillingness to tend to their livestock as reformers thought they should.[62]

Critics of shifting cultivation usually connected it to soil exhaustion, a decline in productivity caused by the spending down of available plant nutrients without any efforts to restore those nutrients (except perhaps through long fallows). But these same critics also implicated shifting cultivation as a cause of soil erosion. For many observers of southern agricultural practices, and for those reformers who sought to change them, the connection between shifting cultivation and soil erosion was a simple one: shifting cultivation was by definition careless farming, and the eroded landscapes that marked the region must have been a result of such carelessness. There is a lot of anecdotal evidence that shifting cultivation did result in erosion in the South. Moreover, exhaustion and erosion were not always easy to tease apart as discrete phenomena. Exhaustion could lead to erosion by destabilizing soils, while subtler forms of sheet erosion were the cause of what farmers called exhaustion. While there is substantial evidence that exhaustion and erosion together characterized the southern landscape of shifting cultivation, the actual material relationship between shifting cultivation and erosion was a complicated one.[63]

There was another characteristic of shifting cultivation and extensive farming that many observers connected to the region's disproportionate problems with soil erosion. With shifting cultivation, a substantial portion of the South was almost always in a state of recent abandonment, and recently abandoned land, absent any sort of stewardship, can be particularly susceptible to erosion if it is not quickly covered with successional vegetation. Stanley Trimble concluded that the "abandonment of fields with little or no vegetative cover was one of the most erosive practices of the European settlers."[64] Writing in 1915, Royal O. Davis noted of abandoned land: "It is then that erosion is most destructive." "The soil is exhausted of organic matter," he continued, "and even before the weeds begin to grow the rains form gullies over the surface. Probably the field will not be put under cultivation again, and in a few years it becomes devastated, without agricultural value, and a menace to the surrounding land."[65] The New Deal sociologist Arthur Raper also wrote about the destructiveness of land abandonment in Greene County, Georgia, during the 1920s, as the cotton plantation system was collapsing there, as some of Jack Delano's photographs, which illustrated Raper's book, made only too clear (see Figures 17 and 18).[66] Many others concurred with the conclusions of Trimble, Davis, and Raper on the dangers of land abandonment.[67] Accelerated erosion on abandoned land was an uneven process across the plantation South. Where some fertility remained and the land was not too sloping, primary ecological succession usually healed over exposed soils quickly. But in places where the soils had lost most of their fertility, where the washing of topsoil had exposed infertile subsoil, or where rilling had grown into gullying, abandonment encouraged continued erosion. The effects of abandonment were at their worst in the early twentieth century, when many began to leave the land for good.

Not everyone has agreed that the link between shifting cultivation and soil erosion was quite so neat. Because lands under shifting cultivation were cleared for relatively brief periods and often contained dead trees, stumps, roots, and other debris that helped hold soil in place, some have argued that this brand of cultivation may have been soil-conserving as compared to clean cultivation in permanent fields. The historian Lorena Walsh, for instance, has suggested that the "hill and hoe culture" of early tobacco farming in the Tidewater region "caused very limited soil erosion, because the patchwork of manmade hills and tree stumps and roots left in the fields retarded both runoff and the leaching of vital plant nutrients."[68] While commentators constantly criticized this approach to tobacco farming as unruly, Walsh found little evidence, observational or otherwise, that it had caused appreciable erosion in the Tidewater region during the colonial era.

Rather than blaming shifting cultivation for erosion, several scholars have argued that it was agrarian reform efforts aimed at ending shifting cultivation that were the true causes of accelerated erosion. Among this group was the late Carville Earle, who, in an effort to critically reconsider what he called "the myth of the southern soil miner," produced an instructive timeline for understanding southern agriculture and its relation to soil erosion over several centuries. Earle argued that it was two waves of "scientific" agricultural reform that produced the worst moments in the region's erosion history, for these reform efforts undermined adaptations that southern farmers and planters had made in reaction to major market downturns, adaptations that, he argued, had

Figure 18. Jack Delano, "Greene County, Georgia." Severe gully erosion on a farm in Greene County. Courtesy of Farm Security Administration / Office of War Information Collection, Library of Congress. This photograph also appeared in Arthur Raper's *Tenants of the Almighty*.

Figure 19. Jack Delano, "Erosion on a Farm West of Greensboro, Greene County, Georgia." This photograph nicely reveals how erosion can continue to be destructive even after land has been abandoned. Courtesy of Farm Security Administration / Office of War Information Collection, Library of Congress. This photograph also appeared in Arthur Raper's *Tenants of the Almighty*.

been soil-conserving. While Earle's argument is more suggestive than conclusive, it is worth examining for its insistence that we attend to place and time in refining generalizations about southern agriculture and its impacts upon the region's soils.[69]

Earle began his examination of the soil miner myth with the earliest settlement of the Chesapeake colonies and the settlers' singular attachment to growing tobacco, which initially proved immensely profitable. When prices dropped in the 1630s, tobacco farmers maintained profitable operations as a result of innovations, such as topping tobacco plants to produce fuller growth, which increased productivity. But then another downturn in prices around 1680 pushed tobacco planters to consolidate their holdings and pursue a different type of innovation: a variation on the forest fallow system described earlier that Earle called "recyclic shifting cultivation." Earle insisted that this system not only restored soil fertility through land rotation but also that "erosion was retarded by the chaotic drainage system of typical tobacco fields, filled as they were with the numerous check dams of stumps, roots, hills, and pits."[70] Refined by the beginning of the eighteenth century, this forest fallow system diffused throughout the Chesapeake colonies over the decades leading up to the American Revolution. But in the late eighteenth century, things went bad, and the culprit, Earle argued, was the "high farming" that agricultural reformers encouraged as an alternative to shifting cultivation. There was a move to continuous cultivation in fields free from the debris, a move that allowed for the use of the substantial plows championed by reformers. There was also a shift to wheat, and to more intensive and sedentary production, just as the reformers had wanted. Avery Craven saw these shifts as positive ones, as frontier practices giving way to a permanent agriculture. But Chesapeake soils began eroding at alarming rates, a trend supported by sedimentation studies of the Chesapeake Bay that have found accelerated deposition beginning in the late eighteenth century and continuing into the early nineteenth century. Contrary to interpretations such as Craven's that blamed shifting cultivation for exhaustion and even erosion and that hailed antebellum reformers as saviors of the region's soils, Earle argued that the reformers themselves were the problem.[71]

This move away from recyclic shifting cultivation and toward "high farming" not only accelerated erosion in the Chesapeake colonies during the late eighteenth century, but according to Earle it also propelled many to abandon their lands and head west onto the Piedmont, a region whose erosive potential was quite different than the Tidewater's. The collision of these reforms with the rise of short staple cotton in the 1790s ushered in what Earle called "a half century of destructive occupance." "The vicious cycle of clearing, planting, abandoning, and migrating to virgin frontier lands," Earle wrote, "was extended from coastal Georgia and South Carolina to the Mississippi."[72] This was a more destructive sort of shifting cultivation, the pattern most people have envisioned when they have condemned southern farmers and planters as soil miners. Earle admitted that it existed across much of the region during this period, but, again, he saw it as the product of reforms that undermined ecologically beneficent forms of shifting cultivation.

After a fit of "destructive occupance," cotton prices crashed in the 1830s and 1840s, and southern farmers and planters again adjusted their land use. In this case, Earle suggested that many quietly settled into a system in which they grew less cotton and began rotating cotton lands with corn intercropped with cowpeas. This system, which seems to have emerged in the Natchez District of Mississippi and spread from there, had two important advantages over the soil mining that had preceded it. First, crop rotation with leguminous cowpeas, which fixed atmospheric nitrogen, restored soils that had been cropped in cotton, allowing farmers and planters to later rotate them back into cotton with good results. Aside from fixing nitrogen, cowpeas provided ground cover that prevented erosion and retained soil moisture. Moreover, they often became swine fodder after the corn harvest, and bringing swine into fields had the added benefit of a modest manuring. This brings us to the new system's second important advantage: raising their own corn, increasing their swine numbers, and harvesting some of their cowpeas allowed farmers and planters to lessen their reliance on food imports from the upper South and to achieve something closer to food self-sufficiency. Again, Earle saw this as a subtle, intelligent, and practical adaptation to depressed market conditions, one that reversed a half-century of bad practices and reduced soil exhaustion and erosion. Here was a system that finally seemed to be allowing southern planters to settle in place, even if it did not fully embrace the reformers' ideal of mixed husbandry. There remain real questions about how widespread these rotational and intercropping practices were. Nonetheless, this system of leguminous intercropping was, for Earle, a compelling example of southerners adapting their agricultural practices to restore soils.[73]

This system of crop rotation lasted into the postbellum years, but by 1880 it was under attack by a new scientific panacea: inexpensive phosphate fertilizers, promoted by the growing number of state departments of agriculture, that led many to abandon cowpea rotation as unnecessary. These phosphate fertilizers were not the first commercial products to promise relief to southern planters and farmers contending with unproductive soils. During the 1840s and 1850s, nitrogen-rich seabird guano, mined mostly from the Chincha Islands off the coast of Peru, entered American markets with much fanfare, promising farmers the ability to import soil fertility onto their lands rather than raising it in situ or migrating in search of it. But Peruvian guano generally was too expensive for cotton farmers and planters to justify its use, and so it had a limited impact on the region. Mostly it was farmers growing high-value crops for urban markets who utilized guano, although guano also found a ready market among wheat farmers in the old Chesapeake tobacco region. Peruvian guano made little headway in the Cotton Belt (though, confusingly, the word "guano" would persist as a generic descriptor of all sorts of commercial fertilizer mixes into the twentieth century). Phosphates, on the other hand, were cheaper than Peruvian guano, particularly after the commercial development of phosphate rock deposits in South Carolina (and later Tennessee and Florida) immediately after the Civil War. Southerners mined and manufactured these deposits into a product known as superphosphates, so-called because treatment with sulfuric acid made the phosphorus in these deposits more soluble and available to plants. When applied to cotton lands, phosphate fertilizers produced a spike in productivity that, according to Earle, lured many farmers and planters away from crop rotation and into cotton maximization. It was not long, however, before the chief limitation of commercial phosphates, their lack of nitrogen, led to declining yields in soils no longer rotated with cowpeas. (Sometimes phosphates were mixed with nitrates from various sources, but the results do not seem to have been particularly satisfying). What followed, Earle argued, was the most destructive period in the history of southern soils, from the 1880s through the 1920s, when the availability of cheap fertilizers encouraged southern farmers, many of them increasingly indebted tenants, to plant more of their land with cotton, even marginal acres, and to use increasing amounts of phosphate fertilizer even as its returns were diminishing. Here Earle's argument meshes well with Stanley Trimble's observation that erosive land use in the South spiked after 1880 and with studies that have found the region's worst pulse of stream sedimentation occurring during the same period. It is not surprising that Providence Canyon gained its modicum of fame at the end of this destructive period.[74]

In making this argument about the role that scientific reform played in producing rather than preventing soil erosion, Carville Earle did not entirely repudiate the myth of the southern soil miner. Rather, he argued that "destructive occupance" in the South occurred in pulses and that such periods alternated with periods of "constructive occupance." In particular, he distinguished between two types of shifting cultivation: the recyclic variety, which he deemed sustainable, and the mine-and-abandon model, which he deemed destructive. The real myth his work attacked, then, was the "linear myth of three centuries of destructive occupance," the "presumption that ten generations of southerners committed one environmental blunder after another."[75] Instead, Earle attempted to reclaim the reputations of white southern farmers and planters at particular moments in the region's agricultural history, and he urged us to be suspicious of the observations and alleged wisdom of "scientific" reformers. While Earle did not entirely redeem shifting cultivation, he did suggest that blanket indictments of southern farming practices, which have been rife in the literature on southern agricultural history, miss important spatial and temporal variations. Finally, he attempted to shift attention away from two other traditional explanations for the destructiveness of southern staple crop agriculture. "The southern soil miner," he concluded, "was the progeny not of plantation economy and slavery (nor their legacies), but of premature epistemologies of scientific agrarian reform."[76] On this point Earle almost certainly overplayed his hand.

Thus far I have discussed the lessons that Providence Canyon might teach us about southern soil erosion in terms of generalized patterns of land use, and without attention to the repressive labor regimes that characterized southern plantation agriculture for most of its history, or to the broader political economy of staple crop production. But one cannot stare for long into Providence Canyon without wondering what roles slavery, tenancy, and international markets played in producing such a spectacle. Indeed, it is impossible to understand the power of the myth of the southern soil miner as it was carried into the New Deal years unless one appreciates the connections that critics made between destructive land use practices and the exploitative labor regimes of the plantation—connections that had more substance than Carville Earle would admit.

What was the impact of slavery on the soils of the South? This is a critical question for understanding the history of southern soil erosion, but it is not an easy one to answer. Before the Civil War, a reliance on slave labor seems to have encouraged an extensive and soil-negligent agriculture in the region. Heavy capital investment in a bonded labor force propelled planters to maximize their profits in the short term by relying on the accrued fertility of fresh lands rather than settling in place and achieving a less-profitable landed permanence. As the economic historian Gavin Wright has argued, antebellum planters' interest in slave property usually trumped their interest in land ownership and stewardship. "Slavery," Wright concluded, "generated a weaker and looser connection between property holders and the land they occupied." Planters who were heavily invested in slave labor and worked in a context of land abundance found it more profitable to practice extensive and shifting cultivation than they did to divert slave labor to the tasks required to sustain soil fertility over the long haul. Since slaves were their primary capital investment, planters were more concerned with how much of a crop they were producing per hand rather than per acre, which meant that slave-based plantation agriculture spread itself over a lot of land in ways that struck many observers as wasteful. As Lewis Gray noted of tobacco and cotton culture in the Old South, "the general disposition in these industries was to economize labor at the expense of land," with the result that "planters frequently undertook a larger acreage than they could attend to." In this way, the region's dedication to extensive and shifting cultivation flowed from the economic logic of slavery, which encouraged land skimming rather than intensive cultivation and conservation practices. Such inattention may have led to accelerated erosion.[77]

Driven by these economic incentives, slaveholders were largely immune to the pleadings of agricultural reformers, even though many came from their class and insisted that reform and slave labor could coexist. Planters resisted practices such as livestock husbandry, manuring, crop rotation, and diversification because all required that they divert cleared land and slave labor from staple crop production. That does not mean that planters never embraced reforms such as mixed husbandry and crop diversification or that they never used slave labor in achieving such ends. The commercially driven, slave-based plantation system did not preclude a model of improvement that devoted more labor and land to soil building. Edmund Ruffin, for one, pleaded with southern planters to adopt a model of improvement whose labor-intensive tasks would have been unimaginable without control over a large bonded labor force. As Jack Temple Kirby noted, "Ruffin and other southern 'improvers' of his generation required slaves to perform the arduously disagreeable labor of landscape manipulation."[78] But such examples were exceptional. Just as it was cheaper to clear an acre of new land than to manure an acre of exhausted land, so it was more cost-efficient to work slaves on fresh lands than to have them rebuild the tired lands left behind. Moreover, the capital tied up in slaves stifled investments in improved livestock, advanced implements, or fertilizers that could have been soil-conserving. As an example, Eugene Genovese disapprovingly cited a plantation in Stewart County that had a fixed capital investment of $42,660 but that had invested only $300 in implements and machinery. Slaveholders with tremendous capital tied up in their labor force felt the profit motive with vigor, and the results showed on the face of the countryside.[79]

Beyond the economics of slavery and the ways in which it drove planter behavior, there is the question of whether the actions of slaves and the inhumane conditions of enslavement worked against soil conservation. For as much as we should heap blame on planters for wreaking havoc on the region's soils, we need to consider the roles that slaves themselves played as the primary workers for much of the region's agricultural land. This is fraught territory because of a long tradition of associating the South's soil-depleting agriculture with the inadequacies of slave labor. Several early historians of antebellum southern agriculture, including Avery Craven, Lewis Gray, and U. B. Phillips, blamed many of its problems, economic and environmental, on the limitations and inflexibility of what they believed was an inferior labor force. They assumed that people of African descent were either incapable of sophisticated farming or in need of civilizational uplift before they could achieve it. As Craven dismissively put it in describing the transition to slave labor in the Chesapeake, "The slave proved more profitable because of his permanence but added nothing in intelligence or interest in his task."[80] This school of thought alleged that an inferior labor force required planters to adopt many of the agricultural practices, such as shifting cultivation, a reliance on primitive tools and techniques, staple crop specialization, and the neglect of livestock husbandry and manuring, that caused environmental damage in the region. Such arguments were long ago discredited, but they appear frequently in the primary sources, and, importantly, they became the basis for several rounds of scholarly revision that have reconceptualized the relationship between slavery and soil degradation in the South.

Working under a system that coerced their labor, often brutally, slaves had few incentives to be stewards of the soil. This seems like a straightforward conclusion, a basic lesson to be learned at Providence Canyon. Writing about what he called "the hard ruthless rape of the land" on the cotton plantations of southwest Georgia, W. E. B. Du Bois noted that "the harder the slaves were driven the more careless and fatal was their farming."[81] Or, as another scholar aptly put it, "Forced labor is as slovenly and poor as the worker, with one eye on the whip, dares to make it."[82] As these comments suggest, slaves may have neglected soils for several reasons: because they had no incentive to care for soil in a system that offered them no rewards for doing so, because they were driven in ways that constantly forced them to choose between expediency and care, and because negligence could function as a form of resistance to an exploitative system. Here the scholarship of Eugene Genovese is critical, for no other scholar so thoroughly pinned the deterioration of southern soils on the institution of slavery. Genovese agreed with the "Phillips-Gray-Craven school," as he called it, on one point: "The slaves worked indifferently," he bluntly noted a half-century ago, concluding that slave labor was characterized by "carelessness and wastefulness."[83] But he did not rest this conclusion on racial or civilizational assumptions about African American inferiority, as had earlier historians. Rather, he argued that slavery as a dehumanizing institution not only bred carelessness among workers—"carelessness" in a literal sense—

but also that carelessness and wastefulness could be important patterns of resistance to the institution. "Side by side with ordinary loafing and mindless labor," he wrote, "went deliberate wastefulness, slowdowns, feigned illnesses, self-inflicted injuries, and the well-known abuse of livestock and equipment." This was willful behavior rooted in the brutality of slavery.[84]

Genovese wed this explanation of slave motivations to an argument about how the economic positions of southern slaveholders encouraged them to take a particular approach to the land. He argued, contra Craven, that not only did exploitative frontier practices persist in the South but that slavery explained, even necessitated, their persistence. The basic economics of slavery contributed to the regional commitment to shifting cultivation and agricultural expansion, but so too did the inefficiencies of coerced slave labor and the active resistance of slaves. Where, for instance, planters and some early historians blamed the region's crude implements on inferior laborers lacking the skills to work with better tools, Genovese suggested that the persistence of crude implements was a product of slaves undermining their masters by breaking tools or using them improperly. Planters and other observers also frequently wrote of the alleged inability of slaves to work with livestock and the poor care that they gave them, suggesting that this explained, or excused, planter failures to practice mixed husbandry. Here too Genovese suggested that the poor treatment that slaves allegedly gave to work animals was a result either of a carelessness bred by the slave system or what he called "vengeful" behavior. Where it existed, the poor treatment of animals by slaves, like the abuse of the soil, was likely an extension of their own mistreatment, of being driven at a pace that made proper care impossible. But slaveholders and those observers sympathetic to their interests often found it convenient to blame slaves for such behavior rather than the system that demanded it.[85] The inferior labor theory provided all sorts of justificatory cover for planters. For Genovese, on the other hand, the system of slavery itself deserved the blame.

As other observers had before him, Genovese also connected the overseer system to land abuse. Since slaves had little incentive to work well and may have intentionally worked poorly when they could get away with it, Genovese reasoned, their labor required constant supervision. The more complex and varied the tasks, the more intensive the supervision slaves required. Again, this was not because slaves could not do diverse and sophisticated work absent supervision but because they would not, at least not when it was serving the agricultural ends of their masters. Soil-depleting staple crop production was easier and cheaper to supervise than more complex and diverse systems and practices that might have been soil-conserving.[86] There were also strong connections between the ways that overseers built their reputations and marketed themselves to planters competing for their services, and cropping practices that damaged the soil. Overseers found their greatest rewards in maximizing staple crop production in the short term, exploiting both labor and land to do so, and since overseers rarely lasted long in their jobs, they were not around to witness the environmental destruction that resulted from driving land and labor in lockstep. The antebellum agricultural reformer John Taylor of Caroline County, Virginia, brilliantly summarized this argument in his classic collection of essays *Arator*. "The farm," Taylor lamented, "is surrendered to a transient overseer, whose salary is increased in proportion as he can impoverish the land. The greatest annual crop, and not the most judicious culture, advances his interest, and establishes his character; and the fees of these land doctors are much higher for killing than for curing. It is common for an industrious overseer, after a few years, to quit a farm on account of the barrenness, occasioned by his own industry."[87]

In Genovese's hands, the connections between the political economy of slavery, the work lives of slaves, and the deterioration of southern soils were clear and direct. "Slavery and the plantation system," he tersely concluded, "led to agricultural methods that depleted the soil." The slave system was not only antithetical to careful soil stewardship, but the structures of the institution stifled efforts at reform.[88] Inhumane coercion made slave labor inefficient, inflexible, and in need of constant and expensive supervision, and it fomented active slave resistance. This in turn encouraged many of the qualities of plantation agriculture that led to the regional exploitation of the soil. For Genovese, slavery seemed to explain it all. When Stanley Trimble looked at the historical patterns of soil erosion on the Piedmont, his evidence supported Genovese's general claim. He found that the distribution of erosive land use in 1860 was "remarkably similar to the distribution of cotton and tobacco production and to the distribution of rural slaves."[89] Following this logic, historians have generally accepted the lesson that slavery played a central role in the degradation of the soils of the South.

Subsequent scholarship, however, has raised important questions about this neat causal linkage. In his examination of the myth of the southern soil miner, for instance, Carville Earle pointed out that the innovation of recyclic shifting cultivation as a constructive approach to managing soil fertility contributed to the shift from indentured servitude to race slavery. In this particular case, he argued, the move to a slave-based agroecological system was soil-*conserving*, not soil-depleting. This led Earle to caution about "the overly simplistic equation of environmental abuse and southern slavery."[90] Moreover, in other crop cultures, rice being a notable one, slavery did not coincide with the destruction of the soil. Slavery, we have learned in recent decades, is not always easy to generalize about, and so too the relationship between the institution and its impact on soils. Slave-based agricultural production varied widely, depending partly on the size of the plantation or farm, the size of the slave labor force working it, the crop culture, and the dictates and sensibility of the landowner. Surely those variations and their impacts upon the soils of the South deserve our attention. Finally, behind this logic is a critical demographic feature of antebellum southern agriculture that I have thus far elided in connecting soil depletion with slavery: a large majority of white southern farmers did not own slaves, yet their practices seem to have contributed to the region's soil erosion as well. While Stanley Trimble's research did suggest a geographical correlation between slavery and soil erosion, one wonders whether there was much difference between the environmental impacts of large plantations worked by scores of slaves, midsized farms with only a handful of slaves, and small farms worked by white yeomen and their families. We need more research on how small-scale farmers in the region treated their soils in comparison to slaveholders and their slaves.[91] Despite the common assumption that slavery was a soil-depleting institution, and the moral satisfaction that comes with that assumption, questions remain about whether slavery can function as a *monocausal* explanation, a determining variable, for soil deterioration in the South. Here Providence Canyon's interpretation might just raise as many questions as it can answer.

Since Genovese's assertive effort to connect slavery and soil depletion several decades ago, there has been remarkably little scholarship on the subject. A subsequent generation of historians has instead taken on the important task of refuting notions that slaves did not bring intelligence, expertise, or innovation to the agricultural labor they performed. We have a growing understanding of the environmental knowledge with which slaves worked and their capacity to not only participate in but also to contribute to the creation of complex agroecological systems in the Americas. Nowhere has this been more fruitfully developed than in the literature on Lowcountry rice culture and in the intriguing though contested thesis that it was essentially an African innovation. Scholars have also elucidated the substantial African botanical legacy in the Atlantic world, and they have added the African continent to formerly Eurocentric conceptions of the Columbian Exchange. They have also demonstrated the remarkable knowledge and skill African slaves displayed in their use of local environments for self-provisioning, not only in terms of small-scale gardening and animal husbandry but also hunting, fishing, and foraging on the plantation margins. Such sophisticated environmental knowledge shaped slave culture and slave efforts to carve out spaces of autonomy within the plantation system.[92]

Now that we have this portrait of slaves as skilled agriculturalists and a historiography emphasizing their environmental agency, a portrait that has come a long way from Genovese's treatment, we ought to return to the relationship between the institution of slavery, slave behavior, and the fate of southern soils. Two examples, both highly speculative, suggest questions scholars might pursue. First, recall Carville Earle's counterintuitive example of how the rise of recyclic shifting cultivation coincided with the transition from indentured servitude to race slavery. Earle saw this shift as planter driven and rooted in the belief that slave labor made better economic sense in a system of shifting cultivation. The historian Lorena Walsh hints at another intriguing possibility. As she notes, "Tobacco was peculiarly suited to long fallow, hoe-and-hill culture, a farming technology totally unfamiliar to Englishmen but one which Africans had mastered over centuries of experience." Scholars have generally seen this culture as a sign of Native American influence, but West Africans too were growing tobacco for local consumption by the late seventeenth century and could have influenced this strategic shift. Walsh argues that the "connection between the evolution of Chesapeake tobacco culture and the arrival of slaves from regions of West Africa where tobacco raising was already well established has so far gone unnoticed, although the evidence is suggestive." In other words, recyclic shifting cultivation, which Earle saw as a clear sign of *planter* innovation, may have been influenced by African slaves and their agricultural practices. It is a supposition that deserves more historical scrutiny.[93]

Another possibility for finding salutary influences that African slaves may have had on the soils of the South involves the humble cowpea. Recall that the cowpea found its way onto cotton plantations in the late antebellum era as a legume that, when it was intercropped with corn, fixed nitrogen, provided soil cover for erosion control, and served as livestock fodder. Antebellum reformers had long praised the cowpea's virtues. John Taylor, a great champion of green manuring, argued that intercropping "peas" with corn over a multiyear period was an effective method for restoring soils, and he suggested that cowpeas could also be an important dietary staple for slaves. "The best kind of pea for the end in view," he wrote, "is white with black eyes."[94] Here he described black-eyed peas, the most recognizable among the cowpea's many kaleidoscopic variations. The other great Virginia reformer of the era, Edmund Ruffin, celebrated cowpeas as "the clover of the South," a comparison to the favored leguminous cover crop of northern reformers.[95] Several historians have cited cowpea intercropping as an example of how white southern cotton planters recognized and worked to combat the exhaustion and erosion of their soils in the late antebellum era, when cotton prices declined. John Hebron Moore wrote that the appearance of cowpeas in their fields "proves that cotton growers of the 'forties and 'fifties, whether small farmers or great slaveowners, deserve more credit as skillful agriculturalists than has since been accorded them."[96] Earle also used the example of cowpea cultivation in his effort to rehabilitate the reputations of southern farmers and planters as agricultural innovators.

The history of the cowpea, however, raises questions about just who deserves the credit for this innovation. Cowpeas are a crop of African origin that crossed the Atlantic on slave ships as one form of slave provisioning during the middle passage. Slaves established cowpeas in their gardens and provisioning grounds in Brazil and the Caribbean, and cowpeas then spread to North America. By the early eighteenth century, as Judith Carney and Richard Nicholas Rosomoff observe, the "African practice of intercropping black-eyed peas with cereals was also observed in the Carolina colony," and during the decades before the American Revolution, the cowpea was a common crop in the Carolina Lowcountry. There are several theories about how Atlantic world planters first encountered cowpeas. Cowpeas may have been established in China and southern Europe prior to North American colonization, so they could have come with Europeans to the Americas. Native Americans were early adopters of the cowpea, so much so that many, including Thomas Jefferson, mistakenly thought that cowpeas were of American origin. But if Carney and Rosomoff's generalization that "plantation owners first encountered African crops in the food fields of their slaves" applies in this case, we might have to amend Moore's boast that the planting of cowpeas "proves that cotton growers . . . deserve more credit as skillful agriculturalists" to allot at least some credit to the slaves who may have helped to move cowpeas into planters' fields. Douglas Helms, a student of Moore, has openly wondered whether Africans were "the most likely transmitters of knowledge about the cowpeas' fertilizing value." Where Genovese's portrait of slaves left little room for such contributions, both of these examples suggest the need to study how African slaves may have contributed, directly or indirectly, to the region's soil-conserving innovations previously claimed by and for white planters.[97]

In considering the impacts of slavery upon southern soils, it is finally worth asking how the sectional crisis and the Civil War played into that story. Slavery was a central cause of that crisis, and some scholars have posited that soil exhaustion and erosion underlay this growing regional schism. Writing in the late 1930s and early 1940s, Rutgers economist William Chandler Bagley made such an argument in his short book *Soil Exhaustion and the Civil War*. For Bagley, southern patterns of soil exhaustion and erosion were "the key to an understanding of the economic significance of the territorial limitation of slavery." Like Genovese, Bagley insisted that the slave system and slave motivations together worked against soil stewardship. He concluded that "slavery, by stimulating soil exhaustion, generates within itself the force of its own extension."[98] Slaveholders were hungry to push slavery into new territory, in other words, precisely because their commitment to slavery and the soil-depleting practices it encouraged demanded continual expansion and fresh lands. If planters were forced to live within the confines of a bounded region whose soils had been systematically depleted, they would have found that the plantation system lost some of its productivity as a result. And because the value of slaves was based in the agricultural profits that they could produce, Bagley also intimated that slaves would have lost value as well. Thus, when northerners who sought to restrict the westward spread of slavery threatened the continued availability of fresh lands, and thus planters' ability to get the most out of their labor, crisis ensued. "There is no doubt," Bagley concluded, "that the northern aim of territorial limitation of slavery presented a very grave threat to the economic security of the slaveholder."[99] While Bagley was careful to insist that soil exhaustion and erosion, which he saw as one and the same problem, were among the many factors that caused the Civil War, he did note that they deserved more attention as a contributing cause. As the geographer Rupert Vance put it in his 1932 masterwork *The Human Geography of the South*, "Lincoln's mild compromise to leave slavery where it existed appeared in reality a threat of extinction."[100]

More recently, John Majewski has added another gloss to this connection between slavery, shifting cultivation, soil exhaustion and erosion, and the expansiveness of southern agriculture. Majewski wondered why many prominent agrarian reformers in the South, those who wanted to end extensive agricultural practices, were also fierce defenders of slavery and the Confederacy. By Bagley's argument, this would appear contradictory. Majewski argues that these Confederate ideologues, a group he calls "extremists," understood full well that a reliance on shifting cultivation was not only territorially expansive but had produced economic stagnation. He suggests that their vision was fundamentally a modernizing one: they wanted to wed slavery to intensive and continuous cultivation as well as to the development of markets, cities, and industry in the region, all through the deployment of agricultural expertise. Majewski is less interested in the question of whether slavery actually could have been reformed in this direction. Scholars such as Genovese had long insisted that it could not have been. Rather, his is an ideological point: a select group of Confederate nationalists did not see the territorial limitation of slavery as tantamount to its extinction, and they deployed agrarian reform to try to loose the chains that had bound slavery to shifting cultivation. The viability of the Confederacy, they thought, necessitated less-destructive approaches to southern soils.[101] Providence Canyon, then, might even function as a Civil War memorial of a sort, a symbol of the role that southern soils played in that conflict.

Strong correlations between erosive land use and exploitative labor regimes did not end with slavery. The systems of sharecropping and tenancy that emerged in the late nineteenth-century South continued to encourage the production of cotton and other staple crops and to discourage soil stewardship. Indeed, the logic of a system that drove tenants into debt, and yeomen into tenancy, precluded adequate attention to soil conservation. As a result, both cotton production and soil erosion peaked in the South during the half-century prior to Providence Canyon's rise to fame in the 1930s, and erosive land use correlated closely with tenancy throughout the region. While the relationship between slavery and soil erosion is a murky one, the relationship between tenancy and erosion is fairly clear.

In the absence of meaningful land reform after the Civil War, freedpeople found their access to landownership limited. As a result, most African Americans who remained on the land as farmers did so as tenants. For those white southerners (and some white northerners) who controlled the majority of agricultural lands of the South, their primary goal was to return these lands to profitable staple crop production, albeit under new labor regimes. Tenant arrangements in the South during the period were complex, ranging from straight cash rentals to varying degrees of sharecropping, arrangements by which freedpeople farmed the lands of others in exchange for a predetermined share of the crop. Sharecropping seemed to offer African American families a modicum of independence, which helps to explain why some freedpeople, in the absence of meaningful land redistribution, initially found sharecropping an appealing alternative to wage labor as well as a welcome retreat from gang labor. White landowners, however, became skilled at turning sharecropping arrangements to their advantage. In some cases, sharecropping came to resemble wage labor, with owners paying tenants a cotton wage rather than tenants paying landowners a cotton rent.

Furnishing merchants also played a novel role in the post–Civil War South. In a region starved for capital, these merchants supplied both small farmers and tenants with seed, fertilizer, and other supplies in exchange for repayment after the harvest. Such a system necessitated that farmers produce cash crops, usually cotton, in order to repay their debts. Crop-lien laws, which gave lenders first rights to the cotton crops of those who had borrowed from them, put creditors in unusually strong positions to collect their debts from farmers and to direct farmer behavior. As a result, tenants often found themselves increasingly indebted to landowners and furnishing merchants (who were sometimes the same people), and many small landowners, white and black, lost their land and were reduced to tenant status. Indeed, tenancy rates skyrocketed in the region, from about a third of all farmers in 1880 to more than half of all farmers in 1930. While the vast majority of black farmers were tenants, by the mid-1930s two-thirds of all tenant families were white.[102] Once they became indebted, tenants found that the incentives to produce bumper crops of cotton to repay their debts only increased in a vicious cycle that usually drove them deeper into debt. Jack Temple Kirby summarized the environmental results: "Metropolitan bankers charged usurious rates to regional and local banks, which passed their costs on to merchants and lending planters, who robbed sharecroppers, who themselves had nothing to rob but the earth."[103]

The collision of tenancy and the crop-lien system with a spreading commercialism in southern agriculture after the Civil War created an era of destructive erosion that was much worse than had occurred in the Old South. Because tenants had no ownership stake in the lands they farmed, their incentives to improve soils were minimal. Moreover, as overseers had been in the days of slavery, tenants were peripatetic, moving in search of better arrangements, and they thus had as their primary concern the short-term rather than long-term productivity of a piece of land. As the authors of the New Deal classic *Report on Economic Conditions of the South* (1938) succinctly put it: "Half of the South's farmers are tenants, many of whom have little interest in preserving the soil they do not own."[104] As the nineteenth century gave way to the twentieth, tenancy in the South encouraged the degradation of the region's soils, which in turn further impoverished the region's farmers.

Those tenants who improved the soils on their rental properties often found that the benefits accrued to the landowner. This was one of the central challenges faced by Jim Crow–era reformers such as George Washington Carver, who urged African American tenants to build their soils by returning organic nutrients to them.[105] As Charles Johnson, Edwin R. Embree, and Will W. Alexander wrote in another New Deal–era classic, *The Collapse of Cotton Tenancy*, "Since the tenant has no legal claims on any improvements he may make, he has no interest in conserving or improving either the land or the buildings. On the contrary, . . . it is to his advantage to rob the soil of its fertility." "As matters now stand," they continued, "the tenant who really works on his place, who labors to restore the soil, who repairs and builds, is merely inviting his landlord to raise the rent."[106] Moreover, tenants rarely had the capital to purchase livestock for the manure they might have provided, and while many would have preferred to raise their own food and to farm in a diversified way, their rental arrangements and growing indebtedness forced them to focus on cotton and other market staples. Tenants often had to go further into debt to purchase food they could have grown, even as rows of cotton came right up to their doorsteps. This period of tenancy thus saw a decline of farm gardens and other forms of self-provisioning by tenants and the rise of a region rife with malnutrition and the diseases that sprang from it, such as pellagra and rickets.[107] On top of all of this, cotton prices dropped consistently from the early 1870s through the 1890s, which further contributed to indebtedness and, ironically, encouraged the planting of more cotton.[108] Tenants in the plantation South thus pushed the land to help deliver them from these increasingly untenable arrangements. When cotton prices rose in the early twentieth century, the region's farmers engaged in a final frenzy of production before the whole system collapsed during the interwar years.

The rise of the commercial fertilizer industry added impetus to this unsustainable system. As we have seen, inexpensive phosphate fertilizers transformed southern farming in the post-Reconstruction era. In Georgia alone, fertilizer use rose from 48,000 tons in 1874 to 478,000 tons in 1900.[109] Indeed, throughout the late nineteenth and early twentieth centuries, most of America's commercial fertilizers were being produced and consumed in the South.[110] The implications were several. First, these phosphate fertilizers provided a pulse of fertility for beleaguered southern soils, and the promising results tempted many to increase their cotton production and abandon efforts they had made at diversification, crop rotation, or other approaches to soil restoration. Second, phosphate fertilizers increased the land base planted with cotton. On the Piedmont, for instance, antebellum cotton production had mostly focused on the lower Piedmont plantation belt, where the growing season was longer and access to river transport better, but in the late nineteenth century cotton spread over the upper Piedmont, which had been a yeoman region of semisubsistence farming.[111] Fertilizers allowed cotton to mature more quickly, which was one reason cotton production increased on the northern edges of its traditional range, while the growth of a rail network in the upper South encouraged yeomen to shift to market production. Fertilizers also allowed farmers and planters to return to lands previously abandoned as unproductive and to farm their lands continuously rather than relying on shifting cultivation. As a result, cotton production continued its westward expansion after the Civil War, into new regions like the Mississippi Delta, eastern Texas, and Oklahoma, but it also pushed the northern bounds of its commercial viability and washed back across previously farmed southeastern lands that were vulnerable to erosion.[112]

Land planted with cotton and the density of settlement in the region, both of which had been kept down by shifting cultivation, began to increase after the Civil War. Between the late 1860s and 1900, Georgia's cropland planted with cotton increased from 30 percent to 47 percent.[113] In turn, the open range came under stress, challenging the livelihoods of herdsmen who had used the lands of others to make a modest living. By the end of the century, laws requiring owners to enclose their stock had become common. Moreover, the capacity of black and white tenants and yeomen to hunt, fish, and gather to augment and diversify their diets declined as the open range did, and southern states passed restrictive hunting and fishing laws, which further contributed to a mounting nutritional and public health crisis in the region.[114] Finally, the purchase of commercial fertilizer was itself a critical ingredient in the growing indebtedness of southern tenants and of the drift of many yeoman smallholders into tenancy. These phosphate fertilizers, then, contributed to a cotton binge in the region that had terrible consequences for the land and the people who worked it.[115] While phosphates only temporarily solved the problem of exhaustion, they exacerbated the problem of erosion. These fertilizers, historian Ted Steinberg has concluded, "made the late nineteenth century the worst period for soil erosion in the South's entire history."[116] He might have added the early twentieth century as well.

The decline of the plantation region by the early twentieth century, and the accelerated degradation of southern soils that went with it, was also a product of the failure of plantation management. As the centralized operations of the antebellum era gave way to dispersed tenant operations, plantation landowners lost their interest in managing their lands, leaving tenants to farm as they pleased, often without the tools or incentives to care for the soil. In many parts of the region, absentee landownership grew as younger generations left their families' plantations for towns and cities, often without installing qualified managers. Even resident landowners seemed to be paying less attention to how tenants cultivated their land. This management failure, in combination with the absence of tenant motivation to care for the soil and the fiscal pressures that encouraged short-term thinking, created a stewardship vacuum throughout the region by the early twentieth century. For Charles Aiken, a leading proponent of this thesis, the spectacular gullies that dotted the plantation crescent were the direct products of this breakdown in land management under tenancy. As he concluded: "Reasons usually cited for the decline of Southern agriculture, including farm tenancy, soil erosion, and the cotton boll weevil, were actually more symptoms than causes of disintegration. Management failure was the underlying basis of the decay."[117]

Several other factors contributed to the decline of the plantation landscape in the early twentieth century and exacerbated the soil erosion crisis. First, there was the boll weevil, which crossed from Mexico into the United States in the 1890s and ate its way across the cotton South. Historians have long ascribed substantial powers to the boll weevil, with many blaming it—or, in some cases, celebrating it—for the collapse of cotton tenancy. In certain places the boll weevil's arrival did result in collapse, perhaps nowhere more so than in Greene County, Georgia, where cotton production dropped precipitously in the years after World War I, from a prewar average of more than 10,000 bales per year to only 350 in 1921.[118] But as historian James Giesen suggests, the boll weevil was a complicated environmental actor. Some of the decline of cotton culture in the region came in anticipation of the weevil's arrival rather than as a result of its depredations, and in many areas farmers quietly went back to cotton, even after celebrating the boll weevil's role in emancipating them from the staple. Willard Range long ago suggested, rightly I think, that the boll weevil was "only the straw that broke the camel's back."[119] The coming of the boll weevil also contributed to another force that undid the tenant system in the region: the migration of people from the land, particularly the migration of black tenants. In Greene County, for instance, the African American population declined by 43 percent and the white population by 23 percent between 1920 and 1930, even before the Great Depression and New Deal crop reduction policies encouraged another round of outmigration.[120] Some moved to the South's towns and cities, but many more moved north and west to seek industrial employment. As tenants left the land, a final pulse of soil erosion and gullying on abandoned farms marked their departures.

Jim Crow segregation itself must bear some of the blame for the region's ravaged landscape. No piece of writing from the era of Providence Canyon's fame better tackled this connection than did Arthur Raper's 1937 essay "Gullies and What They Mean," which appeared in *Social Forces*, the journal founded by Raper's mentor, Howard Odum. Raper served as research director for the Commission on Interracial Cooperation (CIC), founded in the wake of the violent race riots after World War I. While Raper rose to prominence with an influential social scientific study of southern racial violence, *The Tragedy of Lynching* (1933), his major research focus was on sharecropping and tenancy. During the Roosevelt years he published three books on the subject: *Preface to Peasantry* (1936), *Sharecroppers All* (1941), and *Tenants of the Almighty* (1943). Few during the era were as qualified to speak to the connections between racial segregation, tenancy, and soil erosion.

Raper had a more radical take on the meanings of gully erosion than did soil conservationists such as Hugh Hammond Bennett, whose critique of erosion in the South stopped short of tackling its deep-seated racial problems. Gullies, Raper began, "are physical facts with social backgrounds and consequences."[121] They were, he noted, "common throughout the upland stretches of the old slave plantation areas of the South," where "soil depletion has been to gullies as dusk to night." As many others had, Raper blamed gullies on a now-familiar litany of southern habits and patterns: "clean culture cash crops," the credit structures that forced farmers to forego soil conservation, the failure of plantation management, and even the nation's tariff structure. But to these he added the "race factors" that contributed to gullying, for Raper saw tenancy and its environmental destructiveness as rooted in the post-Reconstruction disenfranchisement of the region's African American citizens. Plantation owners, he insisted, had crafted this tenant system not only to return the land to profitable production and to create a pliable labor force but also to use economic dependence to reassert white supremacy and monopolize political power in the region. Plantation owners were thus the "arch gully-makers," in Raper's interpretation, and gullies were environmental emblems of Jim Crow segregation. Part of the tragedy of tenancy, he noted, was how efforts to disempower freedpeople had swept up a much larger segment of the region's farming population. "[T]he early definition worked out to control the ex-slaves," he argued of the tenant system, "still fixes the status of the South's landless farmers, though only one-third of them are Negro." "The gullies are the region's receipts for the 'bargains' the system got out of virgin soil, slavery, and farm tenancy combined," he concluded. But it was human exploitation that most concerned him. As he poignantly suggested, "[T]he plantation system causes gullies by what it does to the man even more than what it does to the land." Thus, preventing gullies required more than a technical fix. "To stop gullies," he concluded, "we shall need to examine our philosophy of human relations, which still emphasizes group achievements less than race and class distinctions." Raper's was a compelling case for seeing gullies as symbols of the need for a "New South" and a "New America," both committed to social and racial equality. Raper's interpretation of gullies, like Thomas Jefferson Flanagan's poetic paean to the "Canyons at Providence," rightly conceptualized gully erosion as an expression of long-standing patterns of environmental injustice in the agricultural South.

Finally, Providence Canyon needs to embody an interpretive message that Hugh Hammond Bennett pointed to in the 1930s, as the plantation South was in decline: "The mills of Manchester and New England must share with the farmers of the South the responsibility for erosion caused by continuous cotton production."[122] Bennett ought to have added the mill complexes of the southern Piedmont as well, which became important cast members in the morality play that was the benighted South. Nonetheless, he was right that the story of cotton transcended the region. Southern cotton went to feed factories and consumer markets around the world, and the producers and consumers of these finished goods were thus implicated not only in the brutal labor systems but also in the eroded landscapes found throughout the South by the 1930s. Southern cotton was the critical raw material for the second industrial revolution, and while the U.S. South dominated world cotton markets before the Civil War, giving it a favored place in the geography of industrial capitalism, after the war it saw growing global competition from places such as India and Egypt, where cotton was produced under similar labor regimes. Cotton production in the South was thus part of a global capitalist system, what the historian Sven Beckert calls an "empire of cotton," and the slave-based plantation system that gave birth to this empire was capitalist to its core. Even as so many of the region's farmers were growing as poor as the lands they farmed, a pattern repeated in other parts of the world where cotton came to dominate, people far beyond the confines of the South were growing fabulously wealthy from cotton-fueled industrialization. Providence Canyon should stand, then, as a monument not only to the land use decisions made by regional planters and farmers, and to a lesser extent to the constrained agency of slaves and tenants, but also to a broader set of actors and market forces that made cotton production so profitable, at least for some. Southern soil erosion was one of the major environmental costs of Anglo-American industrialization and the making of global capitalism.[123]

Providence Canyon State Park ought to be more than an ironic curiosity. It should be a place to meditate on the profound history of human-induced soil erosion in the American South and its diverse causes and effects across the landscape gradients of the region. More than that, it needs to speak to all of the ways in which human motivations and decisions shaped that history, both within and beyond the South. It ought to be a monument to the settler society that produced it and to the market forces that drove those settlers. It ought to be a memorial to the farming practices and strategies that depleted the soils of the South and to the brutal and degrading systems of slavery and tenancy that undeniably contributed to soil erosion across the region. Providence Canyon is indeed a "testament to man and his mistakes," and it sits in a region with no shortage of environmental sins to be washed away. But it also needs to do more. It needs to communicate that conservation cures were sometimes as destructive as the maladies they were designed to alleviate. It needs to raise tough questions about the morally satisfying linkages we have drawn between the exploitation of land and labor. And it needs to embody the extraregional forces that motivated southerners to work the land the way that they did. Providence Canyon embodies all of these realms of human agency. As we will see in the next chapter, that is not all it embodies.

CHAPTER SEVEN

Somewhere between the Grand Canyon and a Sickening Void

WHILE HUMAN ACTIONS clearly triggered the extensive erosion that has left lasting legacies across the plantation South, the southern environment was not merely a passive platform for human transgression. Particular qualities of the southern environment also shaped the region's erosional history. By invoking the environment as a field of causal force, I do not mean to imply that the "nature" of the region determined the course of the South's agricultural history or that environmental patterns and forces alone explain the region's history of accelerated soil erosion. There was a deep vein of environmental determinism in early southern historiography that I do not wish to repeat, a flawed line of reasoning memorably captured in the opening paragraph of U. B. Phillips's *Life and Labor in the Old South*. "Let us begin by discussing the weather," was Phillips's now-famous invitation to his readers, "for it has been the chief agency in making the South distinctive. It fostered the cultivation of staple crops, which promoted the plantation system, which brought the importation of negroes, which not only gave rise to chattel slavery but created a lasting race problem."[1] With audacious facility, Phillips suggested that the South's plantation system and the "lasting race problem" that sprang from it were products of the weather rather than a long chain of human actions and decisions. Southern historians have rightly rejected Phillips's logic. Indeed, few historiographies have done more to emphasize the power of human agency. But this emphasis on human agency has put environmental historians interested in the plantation South in a bind. Most have felt obliged to begin with a refutation of environmental determinism before getting on with the business of explaining how human action in the plantation South was entangled with its material environment.[2] When, in the 1970s, the historian Julius Rubin penned an influential article on the environmental constraints (he called them "non-institutional factors") that limited agricultural progress in the South, he began not with the weather but with an apology for daring to turn scholarly attention back to the material environment as a shaper of southern agricultural history.[3] While it can be dangerous to reintroduce environmental causation into southern agricultural history, then, we cannot make historical sense of southern soil erosion unless we develop soils and the environmental forces that have shaped their history as characters in this story. So let us begin with the soils of the South and tread carefully from there.

While southern soils have a lot of variety, there are several generalizations that one can make about them. First, they are old and highly weathered. Weathering is the process by which soils are formed from underlying bedrock and sediments, so it is a productive process, at least in its early stages. But advanced weathering-stage soils, like those covering much of the Southeast, are sufficiently old that many of the mineral products of the breakdown of parent rock have been leached away. As a result, southeastern soils, most of which are Ultisols, tend to be acidic and mineral deficient, and they have a comparatively low native fertility. Ultisols, one of twelve soil orders, are usually found in hot, humid areas, as heat and humidity accelerate the chemical weathering processes. Heat and humidity also oxidize organic matter rapidly, which works against the accumulation of rich topsoil. As a result, the developmental trajectory of Ultisols is toward acidification and nutrient depletion. In fact, the Ultisols of the Southeast bear a strong resemblance to many tropical soils, which tend to be deeply weathered and prone to exhaustion and erosion if not handled with care. Dan Richter and Daniel Markewitz, who have studied Piedmont soils extensively, suggest that, at the time of settler clearance, these soils were not only acidic but had relatively low levels of organic carbon and nitrogen as well as phosphorus, calcium, magnesium, and potassium. There are some important exceptions to those rules: the Alfisols of the Shenandoah Valley, the bluegrass region of Kentucky, and Tennessee's Nashville Basin, and the Vertisols of the Black Belt prairies of Alabama and Mississippi and the black waxy prairies of Texas do not suffer from acidity and had greater native fertility when settlers first encountered them. There were also pockets of good soils within the Ultisol regions of the South that knowledgeable early settlers quickly identified and settled. But acidity and mineral deficiency marked most of the region, particularly the soils of the Piedmont and the Coastal Plain. The acidic and nutrient-deficient soils of the Southeast are not necessarily hostile to agriculture, but they require lime to offset acidity and various other organic and mineral inputs to increase fertility. Thus, to characterize the soils of the upland South as fertile in their "virgin" condition is to miss these crucial limitations.[4]

Soil acidity was particularly important to southern agriculture into the twentieth century, and its discovery was one of the signature moments in southern environmental history. In the early nineteenth century, the young Edmund Ruffin read and was influenced by the reform ideas of a fellow Vir-

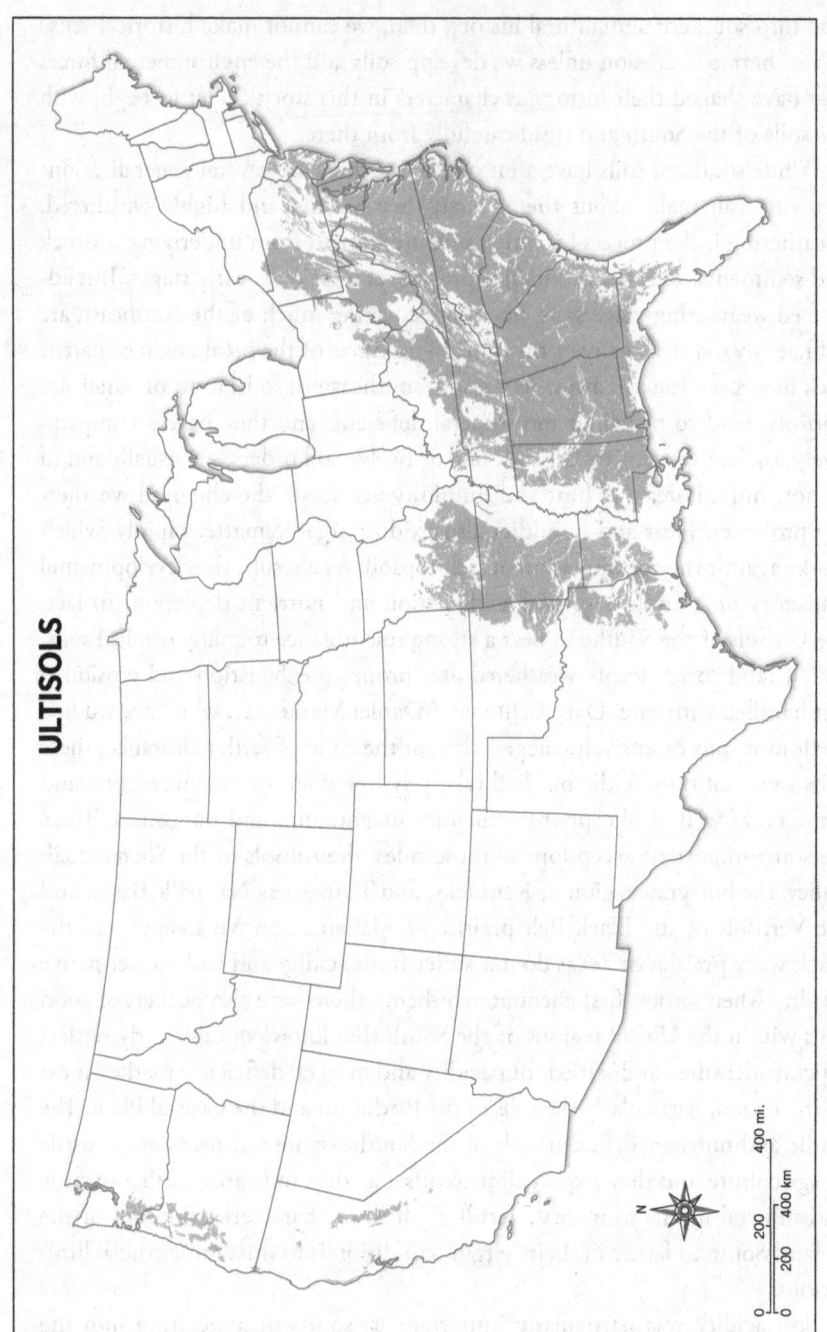

Figure 20. The Ultisols of North America. Ultisols, one of twelve soil orders, are concentrated in the U.S. South. Adapted from a map by the Natural Resources Conservation Service.

ginian, John Taylor, and he adopted Taylor's prescriptions, particularly green manuring and other efforts to return organic material to the soil. But Ruffin met with lackluster results and searched for another explanation for the poor performance of his land. He found it in the work of Humphrey Davy, who had written about the problem of acidity and the solution of applying lime (calcium carbonate) to the soils to reduce their acidity. Ruffin determined that his soils were indeed acidic, a fate that many of the region's other soils shared. With the guidance of one of his slaves, Ruffin learned that there was a ready remedy on his own property: deposits of marl, or fossilized shells, that could be excavated and applied to his fields, arduous labor performed by his slaves. Thus began Ruffin's advocacy for "calcareous manures," his gospel of marl, which he trumpeted all across the acidic South. While marl itself was not much of a fertilizer, its correction of soil acidity allowed plants to take better advantage of existing and applied nutrients while the calcium it contained improved bacterial nitrogen fixation. Marl thus promised to unlock the productivity of soils held captive by acidity and to make the application of organic fertilizers and the planting of green manures more effective. Ruffin's discovery of the acidity of southern soils and his experimentation to correct it earned him a place of primacy in the history of American soil science and agrarian reform. But his gospel of marl attracted a limited number of adherents, partly because it was labor intensive and partly because many did not have the ready marl supplies that Ruffin did. Instead, most of the region's farmers and planters continued with their shifting ways, much to Ruffin's dismay.[5]

As it turns out, the character of most southern soils was an important factor in the region's attachment to shifting cultivation. While high land-to-labor ratios, slavery, and ingrained traditionalism were all important explanatory factors, the acidity and relative nutrient deficiencies of southern soils also contributed to the pervasiveness and longevity of shifting cultivation in the South. On soils that were acidic and nutrient poor, it was difficult to farm continuously without applications of marl and fertilizer. Shifting cultivation worked well as an alternative, and not only because it took advantage of the accrued fertility of forest soils. Shifting cultivators also used fire to convert standing organic matter into soil nutrients that were immediately available to plants, for fire quickly mineralizes the nutrients in organic matter that would otherwise take some time to break down. Moreover, the ash produced by fire culture is alkaline and thus a useful corrective for soil acidity. What shifting cultivators did, then, was use fire to temporarily overcome the limitations of soils that were otherwise acidic and nutrient poor. After several years of cropping, however, the benefits provided by initial fire clearance waned, which then drove cultivators to clear and fire new land.[6]

Once we understand shifting cultivation as a reasonable adaptation to the limitations of many southern soils, we have to rethink the venerable discourse on soil "exhaustion," for it assumes that southern soils were fertile and that shifting cultivation spent down the soil's natural capital. What shifting cultivation spent down was not only the accrued native fertility of the region's topsoil, but more importantly, the added fertility and pH adjustment that resulted from burning the surface vegetation. The discourse on soil exhaustion has created a misapprehension that southern soils were rich to begin with and that farmers and planters profligately squandered that richness, forcing them to migrate to fresh land. In fact, shifting cultivation was an adaptive use of fire culture to compensate for soils that were otherwise challenging to work continuously without fertilizers and lime. Shifting cultivation had its ecological consequences for the region's forests as well as for its soils, but it was also an effective way of contending with the limitations of Ultisols in a situation where land was abundant and labor limited.[7]

The qualities of southern soils also made the prescriptions of antebellum agricultural reformers difficult to put into practice. Many of the pasture grasses and legumes favored by reformers in other regions as both high-quality livestock fodder and green manure—crops such as timothy, red clover, and alfalfa—struggled in the South's acidic soils and with the heat and occasional droughtiness of deep southern summers.[8] "The South has never been a good grass and hay country," the geographer Rupert Vance noted of the region in the 1930s, echoing a long line of farmers who knew this oft-heard regional generalization to be true.[9] The challenges of improving pastures with the traditional suite of pasture grasses and legumes in turn meant that it was difficult for most southern farmers and planters to practice intensive livestock husbandry or raise the high-quality manure needed to maintain fertile soils and sustain continuous cultivation. Moreover, the warm southern climate meant that farmers, planters, and herders could get away with letting their stock roam the open range year-round. They did not need to devote croplands to raising winter feed, to build expensive barns to store that feed and shelter stock, or to bring in their animals seasonally to protect them from winter weather, an annual rite in colder climates that made the production and collection of manure much easier. In that sense, soil quality and climate together steered southern farmers and planters away from intensive grass husbandry and the production of high-quality manure and toward the open range practices that accompanied shifting cultivation. These were not the only paths southern farmers and planters might have taken, but shifting cultivation and the open range were sensible strategies to meet the goals and desires of planters under the environmental circumstances.

Despite these environmental constraints, a few southern reformers were still great advocates of grass husbandry, and they devoted considerable effort to solving the region's "grass problem." One of these advocates was the Reverend Charles Wallace Howard of Kingston, Georgia, who railed against the soil-killing ways of the region's planters and argued for the importance of grass culture as a remedy.[10] His *Manual of the Cultivation of the Grasses and Forage Plants at the South* (1873) crystallized several decades of his grass boosterism.[11] While Howard was sanguine about the possibilities of adapting traditional pasture grasses and forage crops to the region, less optimistic southern planters and farmers explored alternatives to the favored temperate zone grasses and legumes. Not only did they substitute legumes such as cowpeas and lespedeza, they also experimented with both native and imported humid-zone grasses. By the late nineteenth century, the USDA added its expertise to solving the region's grass problems, with bulletins such as George Vasey's *Grasses of the South* (1887). Together southern reformers and USDA scientists found some intriguing possibilities, such as Bermuda grass and Johnson grass, both of which showed promise as pasture grasses that could thrive in the region's soils and endure its hot summers.[12] The southern grass problem, then, was tricky but not insurmountable.

The champions of grass had to get over another high hurdle, however. For cotton farmers and planters, much of the heaviest labor during the crop cycle went to weeding, and invasive grasses were among their most feared foes. As Charles Wallace Howard succinctly put it, "The deadliest enemy of cotton is grass."[13] Many a cotton planter rued a rainy stretch at a critical point in the growing season that produced an infestation of grasses, and many of the grasses compatible with the region's soils and climatic conditions, such as Bermuda, could be particularly troublesome pests. As a result of their distaste for grasses as weeds, cotton farmers and planters rejected suggestions that they cultivate grasses on their properties so as to enclose livestock and produce manure. When, in the late 1840s, the *Southern Cultivator* advocated the planting of Bermuda grass as a path to mixed husbandry, readers responded by pointing out that some Georgia Piedmont plantations had been abandoned in recent years because of Bermuda grass infestation.[14] In another notable case, a Mississippi planter sued a neighbor who attempted to cultivate grass, charging that such an act was an imminent threat to his cotton crop. The suggestion made by some reformers that planters and farmers rotate their fields between cotton and pasture struck many as ludicrous, for it meant planting cotton on lands where residual grasses would spring up and create major weeding work. As Vance noted, "[c]otton and corn culture, agriculture without grass, have in the South met hardy grasses that persist in spreading and resist uprooting. The

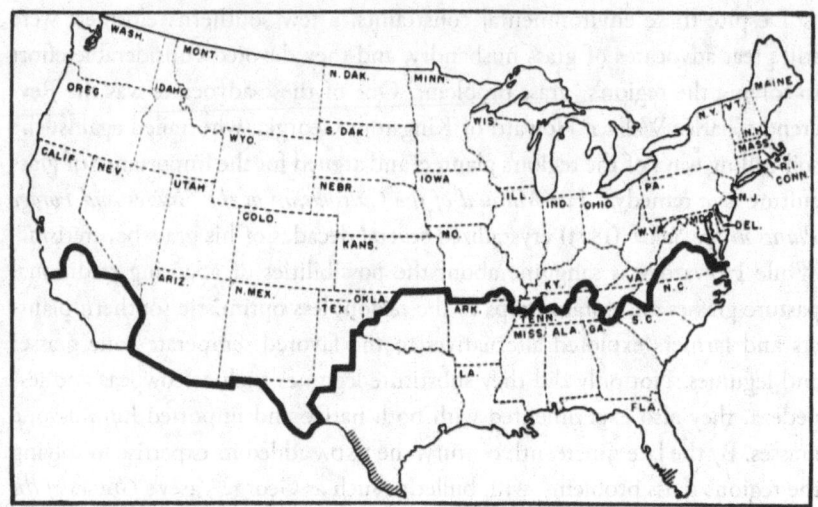

Figure 21. Line of cattle tick infestation in 1906. From *The Yearbook of the United States Department of Agriculture*, 1942, 575.

result is that southern farmers must wage war on grass."[15] Not until the middle of the twentieth century did the USDA and its cooperative programs succeed in breeding pasture grasses adapted to southern environmental conditions, most famously when pioneering agrostologist Glenn Burton produced a Coastal Bermuda that proved an ideal pasture grass for the region. As Willard Range noted, these efforts "forced upon farmers the revolutionary idea that grass was their friend rather than their enemy." Before this point, however, advocacy for mixed husbandry in the South, and the soil-building it might have allowed, struggled against an understandable regional chauvinism against grass.[16]

Another environmental constraint to enclosing livestock in the South and using manure to build soil fertility was a suite of livestock diseases that thrived in the region. While southerners raised a substantial number of hogs on the open range, which provided them with much of their meat, diseases and parasites such as hog cholera and swine kidney worm thwarted intensive hog production in the region until fairly recently. More important to the potential for mixed husbandry, though, was the presence of babesiosis, a cattle disease better known as Texas fever, which powerfully shaped the southern livestock economy. Two species of cattle ticks spread the protozoa responsible for Texas fever, and the presence of the tick vectors defined the disease's geographic extent. As the historian Claire Strom notes, the cattle ticks of the South, "like the cotton plant, required two hundred or more frost free days per year for survival." Not surprisingly, then, the range of tick habitat was essentially coterminous with the cotton South, and Texas fever was thus a distinctively southern environmental constraint.[17]

While outside observers and regional reformers constantly bemoaned the poor state of southern livestock, they rarely accounted for the roles that tick infestation and Texas fever played in producing those poor conditions and limiting intensive livestock husbandry. Indeed, it was not until the end of the nineteenth century that scientists unraveled the etiology of the disease. Importantly, free-ranging cattle in the lower South developed immunity to Texas fever, which meant that the disease was not a direct threat to most southern stock, though continual tick infestation limited weight gain, diminished milk production, and damaged hides. Texas fever and the ticks that spread it thus help to explain the region's reputation for scrawny cattle. Because tick infestation and Texas fever diminished the commercial value of southern cattle, it made less sense for farmers and planters to devote labor to raising cattle when cattle could get along fine on the open range. For the most part, the infected but immune cattle of the lower South met local subsistence needs or supplied regional markets with a second-rate product that could be raised with minimal effort. But southern cattle infected with Texas fever could not mix with northern and western cattle without infecting and killing them. This had two important implications. First, farmers and herders in the tick-infested portions of the South could not send their cattle to the North or Midwest to have them fattened on corn and sold in those regions' growing urban markets. Second, southerners who hoped to improve their stock, and thus their manure production, by importing purebred cattle from other regions found that those imports rarely survived. As Willard Range noted, "imported cattle died like flies" when brought into Georgia.[18] Cattle ticks and Texas fever, then, were substantial environmental constraints on southern livestock husbandry, and their presence helps to explain the reticence of southern farmers and planters to embrace reformist prescriptions that might have built soil fertility and allowed for continuous cultivation.[19]

Some historians have suggested that the persistence of shifting cultivation in the South can be explained almost entirely by looking at these environmental constraints.[20] John Majewski, for one, correlated low percentages of improved land within farms (his statistical surrogate for shifting cultivation) with the presence of Ultisols and livestock diseases. Indeed, he showed that the percentage of improved land within farms in the fertile Alfisol regions of the upper South, most of them above the tick line, was much higher than it was in those sections of the South dominated by Ultisols. Where the soils allowed it, in other words, southern farmers were willing to practice continuous cultivation.[21] But other historians have been less sure that these environmental factors precluded careful and land-intensive modes of agricultural production. Steven Stoll, for one, instructs that the "American South was not equatorial Africa," and that "compelling as they are, environmental explanations cannot account for a political and economic regime that depended on the waste of land."[22] As Stoll intimates, while these environmental constraints definitely influenced agriculture in South into the early twentieth century, so too did the commercial goals of the region's farmers and planters, how they chose to organize agricultural labor, and their decisions to stick with the growing of market staples such as tobacco and cotton. Seeing shifting cultivation as a necessary adaptation to southern environmental constraints risks obscuring the political, social, cultural, and economic goals of the region's farmers and planters, the very goals that defined certain environmental qualities as constraints.[23] Nonetheless, and given these goals, the qualities of southern soils, the challenges they posed to grass husbandry, and the prevalence of certain livestock diseases did shape the regional commitment to shifting cultivation as a strategy for managing soil fertility. To the extent that the persistence of shifting cultivation and the failure to adopt the prescriptions of reformers resulted in soil erosion in the region, these environmental factors need to be considered as part of the cause.

Acidity and nutrient deficiency were not the only challenging qualities of southern soils. As Rupert Vance noted, "The South is more susceptible to erosion than any other section of the country."[24] A good way to explain the susceptibility of southern soils to erosion is to employ a standard formula used by soil scientists to predict soil loss: the Universal Soil Loss Equation (USLE). The USLE states that $A=R+K+LS+C+P$, where A is total soil loss, R is the rainfall erosivity index, K is the soil erodibility factor, LS is the topographic factor or the combination of length and slope, C is the cropping or vegetative cover factor, and P is the conservation practice factor.[25] The beauty of this formula is how it lays out the variables affecting soil erosion. It distinguishes human or cultural from nonhuman or environmental factors, while insisting that they also be seen together as part of a single framework. In chapter 6, we saw how particular crops, agricultural techniques, cropping and labor practices, and conservation regimes (basically, the C and P factors) affected soil erosion in the South, but we also need to account for the R, K, and LS factors.

Rainfall erosivity, which measures not only total rainfall but also the energy with which it hits the ground, is critical to erosion on cultivated land, and rainfall erosivity is higher in the Southeast than in any other region of the United States. East of the Mississippi, rainfall erosivity increases as one moves from north to south, and as one heads west these north–south gradients dip toward the Gulf of Mexico. Rainfall in the Southeast is not only heavy, it is more intense than in other regions of the United States, and it is concentrated in spring and summer, when crop-bearing fields are particularly prone to soil erosion.[26] All other USLE variables being equal (and, as we will see, they were not), a cultivated field in the lower South would experience significantly more erosion, perhaps two to three times more, than an identically cultivated field in New England or the upper Midwest because of the differential in rainfall erosivity alone.[27] Moreover, because the ground rarely freezes in the South and southern soils are rarely protected in the winter by a blanket of snow, rainfall erosivity is a year-round process there. As Arthur Hall noted, "Mild winters lacking in snowfall but accompanied by rain and variations in freezing and thawing cause soil erosion to continue throughout the year."[28]

A second important aspect of rainfall erosivity is the differential between the amount of rain that falls in a given period and the infiltration capacities of particular soils. Soil scientists categorize soils into different hydrologic groups—A, B, C, D—with A soils having the highest infiltration capacities (8–12 mm/hr) and D soils having the worst (< 1 mm/hr). Soils with high clay content tend to have lower infiltration capacities, which suggests that many southern soils may not have performed well in comparison to other soils. But even assuming that southern soils were in the A range, the region's high rainfall rates meant that there was a greater differential between the amount of rain that falls in a given period of time and the amount that even a good soil can absorb. That excess necessarily drains as surface flow. Finally, the removal of forest cover in the South and its replacement with crops significantly reduced evapotranspiration rates, or the rates at which plants move water from the ground into the atmosphere, which in turn would have meant a substantial increase of surface flow on croplands. One of the results of these high differentials between rainfall, infiltration, and evapotranspiration rates was that soil conservation practices such as contour plowing and terracing, which worked in regions with more moderate rainfall rates and higher infiltration capacities, often failed in the South, where soils simply could not absorb the excess rainfall.[29] A high rainfall erosivity index and rainfall rates well in excess of soil infiltration and evapotranspiration capacities played key roles in accelerating southern soil erosion on cleared lands.

Many southern soils also have a relatively high K factor, which measures the cohesiveness of soil types, their resistance to being dislodged and transported by rain, and other physical soil and subsoil conditions that create erosion tendencies. This is partly a result of the fact that many southern topsoils were, at the time of clearance and settlement, poor in organic materials. As the soil scientist David Montgomery notes, "Higher organic matter content inhibits erosion because soil organic matter binds soil particles together, gener-

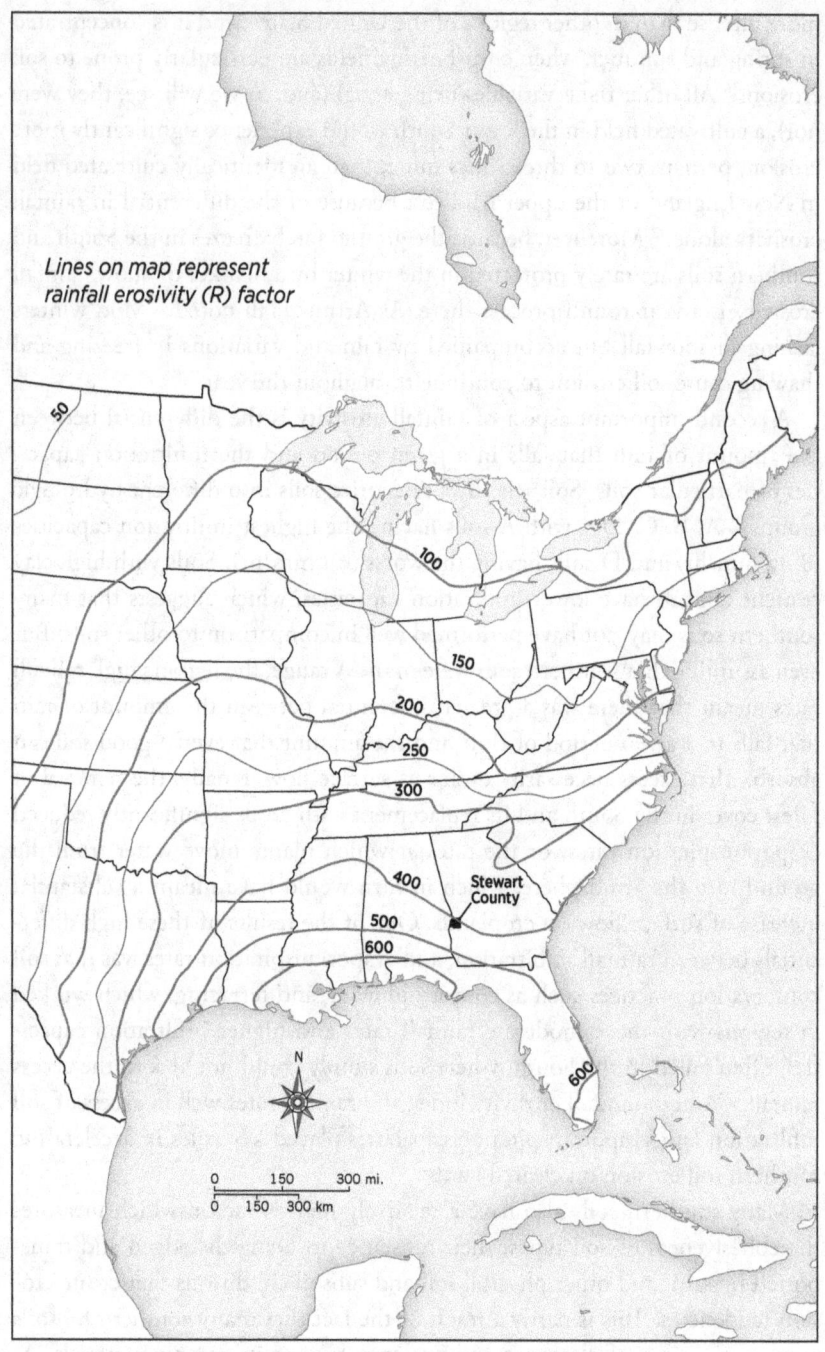

Figure 22. Rainfall erosivity map of the eastern United States. Adapted from a map by the Agricultural Research Service, USDA.

ating aggregates that resist erosion."[30] In the South, relatively low levels of organic matter made soils susceptible to erosion, a factor that was exacerbated by farming practices that removed nutrients without returning sufficient organic materials to the soil. Moreover, the high clay content of many of the region's subsoils meant that they resisted the infiltration of water and produced more surface flow. Rainfall would penetrate the shallow topsoil and then hit impervious clay subsoil, at which point the water would head downhill, carrying the topsoil with it. Shallow plowing atop soils such as these could be particularly destructive because it often loosened the surface layer while creating a water-resistant plow line in the subsoil. In some places, intractable clay subsoils produced what southerners called "galls" or "galled land," areas of clay hardpan from which topsoils had washed and on which almost nothing could grow. In other places, concentrated surface flow ate into the region's subsoils. Again there is substantial variation in the region's soils and their susceptibility to erosion, but much of the agricultural work of the plantation South was accomplished in soils that were more erosion prone than those in other agricultural regions. As importantly, much southern agriculture occurred on lands with considerable slope and thus a high LS factor. Farming erodible soils on sloping lands in a region with high rainfall intensity was thus a recipe for accelerated erosion.[31]

These dynamics have been documented most extensively for the Piedmont. When Stanley Trimble completed his study of Piedmont erosion more than four decades ago, he chose an apt metaphor for the region. He argued that the Piedmont, with its combination of "steep slopes, erodible and deeply weathered soil, and extremely intense rainfall," was an "erosional tinderbox." Undisturbed, the Piedmont did not experience extensive erosion because it supported dense vegetation that protected its soils. But once humans cleared that vegetation, they encountered a combustible situation. Trimble noted that the dominant Piedmont soils had fairly high K factors, which contributed to the historical erosion in the region. But more important to the total volume of eroded soil and subsoil was how deeply weathered Piedmont soils were. A layer of saprolite, or weathered and rotten rock, often underlies the Piedmont's soil horizons, which makes the region prone to deep gullying. While Trimble meted out substantial blame to southern planters and farmers as well as the systems of slavery and tenancy, which he correlated closely with the region's worst erosion, he also admitted that he was not sure "that any other farmers could have grown the same crops on the Southern Piedmont without many of the same results as we have seen." The same cropping and conservation practices that characterized the plantation South, had they been utilized in an environment with far less latent erosive energy, would have produced much less erosion. The larger lesson is this: different soils in different topographic and climatic settings will react quite differently to the same human actions, and histories of human-induced erosion must account for these variations. "Geography is not destiny," Trimble concluded, "but it is a powerful context."[32]

Together, then, heat and humidity, rainfall patterns, soil erodibility, and uneven topography made southern soils particularly prone to erosion when triggered by human action. These environmental qualities must share the stage with human land use practices and political economy in explaining the extent of historical soil loss in the South. Of course, little soil erosion would have occurred if settlers had not cleared and farmed large portions of the region, and much less soil would have been lost if farmers and planters had adapted their goals and practices to the peculiar environmental factors that made the plantation South into an "erosional tinderbox." Certainly one of the great failures of planters and farmers in this region, and of settler agriculture in many other regions, was their inability to square the economic logic that drove them with the particular environmental conditions that they encountered. If one simply wants to explain what caused soil erosion in the South, then human actions deserve the blame, just as one blames arson on the igniter and not the fuel. But if one wants to explain why soil loss was *so much worse* in the South than in other regions—if one is interested in making comparative regional generalizations of the sort that observers of southern agriculture have long been making, including those New Deal–era soil conservationists and environmental reformers who made Providence Canyon stand for a region-wide pathology—environmental factors may have been just as important as human ones in the South's extreme patterns of soil erosion.

The argument of this chapter thus far has been that while human activities clearly precipitated the rampant soil erosion across the plantation South, the comparative magnitude of southern erosion is as much a product of environmental factors as it is of distinctive regional agricultural and soil management practices. But I also need to contend with a question that challenges my efforts to make Providence Canyon interpretively representative of larger southern patterns, cultural and environmental: why were the gullies in Stewart County *so much worse* than the landscapes of erosion found in most of the rest of the plantation South? To answer this question requires moving beyond arguments for southern distinctiveness, cultural and environmental, and giving voice to the peculiar *local* conditions that have made Providence Canyon so unusual, spectacular, and even park worthy.

The first obvious question to ask is whether farming practices in Stewart County were any more destructive than practices elsewhere in the region.

Did the spectacular gullying there correspond with particularly egregious land use practices? The answer appears to be no. Stewart County's agricultural history was fairly representative of the plantation South as a whole. Almost from the beginning of frontier settlement, Stewart was a cotton county, an outpost of an expansive plantation economy reliant upon slave labor and distant markets. By 1850, just a quarter-century after settler agriculture first came to the region, Stewart County produced 19,165 bales of cotton, which ranked it second among all Georgia counties. Slavery came to the county quickly. By 1840, only a decade after its creation, Stewart County had 4,759 slaves out of a total population of 12,933. By 1860, the slave population had surpassed the free population.[33] The county's attachment to shifting cultivation is less clear. While the 1850 census revealed that only 38 percent of the county's land in farms was improved, a figure roughly in line with John Majewski's portrait of the region (though a fair bit higher than the 28 percent for the state as a whole and 29.5 percent for the cotton South), the 1860 census showed that almost 52 percent of the county's land in farms was improved, and while that figure dipped over the next several decades, it was still 45 percent as of 1880. Land use in the county may have been more intensive than elsewhere in the region.[34]

Cotton planting increased in Stewart County in the years after the Civil War, as it did elsewhere in the plantation South. The peak year was 1890, when county residents planted more than fifty-three thousand acres. While the county's early farmers generally eschewed soil-conserving practices, terracing had become widespread in the county by the end of the nineteenth century, though poorly designed and controlled terrace outlets seem to have done as much to cause gully erosion in the county as to prevent it.[35] Again, this was a fairly typical story. The county's economy continued to be driven by cotton production into the early twentieth century, and the poorer residents of Stewart County experienced the same drift into tenancy that small farmers in much of the rest of the region did. In 1880, 64 percent of the county's farms were owner operated, but by 1890 ownership had dropped to 29 percent and by 1913 it was only 19 percent.[36] In 1911, Georgia farmers produced 2.8 million bales of cotton, the most in the state's history, with production centered in the western Piedmont and upper Coastal Plain.[37] That year, Stewart County produced 15,540 bales of cotton on 43,762 acres. But several years later, cotton production began a precipitous decline. By 1920, only about twenty thousand acres were planted with cotton in the county, and remarkably those acres produced only about four thousand bales. Yields per acre picked up slightly over the next quarter-century, but acreage in cotton declined until 1945, when fewer than two thousand acres were in cotton.[38] During the first half of the twentieth century, the county's population declined as well, from 15,856 in 1900 to only 7,371 by 1960.[39] By 2012, county population was down to about six thousand, and, as we have already seen, the county is now more than 90 percent forested.[40] In the century after its initial frontier settlement, then, Stewart County witnessed patterns of land use fairly typical of the plantation South, and by the 1930s Providence Canyon spoke eloquently, if in exaggerated tones, of the environmental consequences. Land use practices certainly help to explain why erosion was a problem in Stewart County, but we will have to look elsewhere to discover why gullying was so extreme.

The gullies of Stewart County did have something to do with locally heightened manifestations of the environmental forces and characteristics that contributed to soil erosion throughout the region. The undulating topography of the county and the significant elevation differences between its uplands and its Chattahoochee bottomlands certainly contributed to the extreme erosion to be found there. As David Long and the other authors of the USDA's 1913 "Soil Survey of Stewart County, Georgia" noted: "The eroding forces have been especially active on account of the great differences in elevation between that part of the county that represents the original upland and the Chattahoochee bottoms."[41] Rainfall erosivity is also extreme in Stewart County, for there are few places in the South that combine such significant relief with the high rainfall erosivity scores of the southeastern Coastal Plain. Providence Canyon would not have developed into the supreme specimen of gully formation that it did, and much of the rest of the area would not have gullied so badly, had those two factors not been unusually strong in this locale. But these enhanced environmental forces cannot wholly explain the county's enormous gullies either. We need to probe a bit deeper.

The most important factor in explaining Stewart County's extreme gullying was the unusual subsurface geology in the area, for under conditions of disturbance, portions of the county could be catastrophically unstable. Providence Canyon's stratigraphy, revealed so helpfully in the canyon's walls, is the product of the deposition of marine sediments that occurred between eighty-five and sixty-five million years ago, when the area was alternately underwater and exposed. At different moments in the deep past, sea level extended all the way to the Coastal Plain–Piedmont border, though it fluctuated so that sometimes the Providence Canyon area was covered by water and sometimes it was not. Looking at the canyon walls, one can see two distinctive layers (figure 23). The top layer, known as the Clayton Formation, is composed of reddish sandy clay, likely washed down from the Piedmont in an earlier geological era. Beneath the Clayton Formation is a deeper layer, extending to well over one hundred feet deep in places, known as the Providence Formation (named by the geolo-

Figure 23. This untitled photograph, likely by Arthur Rothstein, clearly shows the darker Clayton formation sitting atop the lighter Providence sands. Courtesy of the Farm Security Administration / Office of War Information Photograph Collection, Library of Congress.

gists Veatch and Stephenson after its most famous exposure), which is made up of unconsolidated sand with scattered silt and clay deposits. These Providence sands are largely structureless, but they had long been held in place by the Clayton layer that sat atop them. Beneath the Providence sands is another formation called the Ripley Formation, which forms the base of the canyons and is resistant to erosion. These layers, and several others beneath them, gently incline in a formation known as a cuesta, which terminates in a north-northwest facing ridge of more than six hundred feet in elevation. Most of the gullies in the area formed on the cuesta's steep front, where the Providence sands were highly susceptible to erosion.[42]

When we understand these local geological conditions, we can reconstruct, in a speculative way, the county's erosion history. When the first white settlers and black slaves arrived in Stewart County and began farming, they found a thin layer of topsoil atop the clay subsoils of the Clayton Formation. Clearing the forest and plowing the land loosened the topsoil in the hilly western part of the county, where much early agriculture focused, and crops removed nutrients and organic materials, making the soils less stable. Land clearance also altered the county's hydrological cycle, increasing runoff substantially. When rains came, they saturated the thin topsoils and then hit the less pervious Clayton clays. The increased runoff then carried away the topsoil and started to wear away at the Clayton clays underneath. This was a typical southern story thus far. But once surface runoff breached the Clayton clays, reaching the Providence Formation underneath, the erosive power of water encountered a deep and structurally unstable layer of sediment that was susceptible to dramatic gullying. As we have already seen, gullies began forming in the county not long after initial settlement, and perhaps even before commercial cotton growing came to dominate the county's agriculture.[43]

There were two main erosional processes at work in Stewart County, each of which contributed to Providence Canyon's unique features. The first was surface or overland flow, by which rainwater moving across the surface of the land dislodged and transported soils. This is the process most people envision when they think of erosion. When surface flow became sufficiently concentrated and forceful, it could breach the Clayton Formation and expose the unstable Providence sands below. When flowing water encountered the Providence sands, they washed easily, and the depth of the Providence formation meant that gullies could deepen and lengthen quickly. Once a gully began to form, concentrated surface flow propelled its headward migration, or its continual development upslope. Where surface water breached the Clayton Formation, an overfall with a plunge pool could develop, which would eat away at the sands below. Moreover, some of the water flowing over the lip of the Clayton clays at the headwall would cling to the lip as a result of surface tension, flow underneath the lip, and hollow out the Providence sands, eroding them back until there was a cave-like space underneath the surface horizon of Clayton clay. When enough of the Providence sands had been washed out from underneath the Clayton clays at the gully head as a result of this back trickling, the remaining soils above the hollowed-out space caved in, a process known as gully-head caving. Then the dynamic would begin anew. Surface flow produced by a heavy rainfall could result in dramatic caving incidents.[44] By such a process, gullies worked their way upslope until they hit a ridgeline that pulled water away from them.

Surface flow, however, was not the most important erosive force in Stewart County. Geological surveyor D. Hoye Eargle noted of the area, "Only at a few places, where an unusual amount of surface water has been directed into gully heads, was the erosion by scour and plunge or overfalls found to be the chief

Figure 24. "Erosion. Stewart County, Georgia." This photo by Arthur Rothstein shows a path of concentrated surface flow leading to the edge of a gully. Courtesy of the Farm Security Administration / Office of War Information Photograph Collection, Library of Congress.

agent of the erosion of these gullies." Instead, a second process was at work, this one more mysterious and distinctive to the place, which occurred beneath the surface and out of view. Even in the absence of concentrated surface flow, rainfall could infiltrate the Clayton clay layer and percolate down through the Providence sands. The removal of trees, with their deep taproots and substantial thirst, increased the amount of water that did so. This water would move downward until it hit an impermeable barrier. Sometimes that was the Ripley Formation at the bottom of the Providence sands. Water would then flow along the seam between the Providence and Ripley Formations, destabilizing the looser Providence sands above the seam. In areas of the county where the boundary between the Providence and Ripley Formations was exposed, perhaps as a result of the landscape's dissection by streams, water often emerged as seepages or springs. As Eargle noted, "The relatively slow but constant erosion by seepage from springs at the base of the loose Providence sand and near the base of relatively steep slopes is certainly one of the factors that produced the spectacular gullies of the Providence Canyons type."[45]

Frequent lenses of clay called kaolin also exist within the Providence Formation and played an important role in the county's spectacular erosion. Kaolin has its own fascinating history in the region. It has been mined extensively throughout Georgia over the last three quarters of a century, with the heaviest mining activity occurring in a Kaolin belt to the northeast of Stewart County. Kaolin has numerous uses, from providing magazine paper with its glossy sheen to soothing upset stomachs and treating diarrhea—for a long time it was the main ingredient in Kaopectate. Kaolin mining has been a source of tremendous controversy in Georgia because of the ways in which large mining corporations have purchased subsurface mineral rights from poor rural landowners and then exercised them in devastating ways. In Stewart County, kaolin was not abundant enough to support a mining industry, but it did create another form of havoc. When water percolating downward through the Providence sands hit impermeable lenses of kaolin, it moved sideways and downhill, initiating a process called piping or pipe flow, whereby laterally moving water washed away sand to create pipe-like tunnels beneath the surface. When enough of the sand had been eaten away by such subsurface flow, the land above it would slump and slide down hill, resulting in a phenomenon known as mass wasting. By this process, large amounts of sediment collapsed into the county's expanding gullies. These caving events were as dramatic as headwall caving, but they occurred without any obvious signs of water flow having caused the erosion. Such dramatic caving from mass wasting, along with the headwall caving, was what led locals to call the area "Providence Caves."[46] To paraphrase Stuart Chase's CCC guide, sometimes she did go in an acre at a time.

Once we understand the county's unusual geological profile and the erosional processes at work in creating its spectacular gullies, some of the local lore about the place seems less fanciful. Whether Providence Canyon or any of the county's other large gullies were actually the product of a dripping barn roof, a woman throwing out her wash water repeatedly in the same place, cows habitually following the same path down to a spring, or even Creek Indians walking to the Chattahoochee is difficult to determine, but they certainly could have been. A persistently dripping barn roof might have produced just the sort of erosive wear necessary to breach the Clayton clays and send water down into the unstable Providence sands, initiating gullying and mass wasting. Whether such origin stories are true or apocryphal, their unifying moral—that the county's remarkable erosion was a freak accident rather than the predictable result of poor land use practices over generations—becomes plausible. The land could, and often did, collapse quite suddenly and without obvious human cause. There certainly were cases where local residents could see gully erosion as a logical extension of visible surface erosion on agricultural lands. But there were other cases where gullies appeared, deepened, and widened without any warning signs of their incipience. Recall Jonathan Daniels's 1938 observation that the land surrounding the gullies "did not seem to an unpracticed eye badly worn, soils about to collapse in dramatic canyoning." For Daniels, there was no visual clue in the fields surrounding Providence Canyon of a slippery slope proceeding from poor farming to deep gullying. Rather, he saw gullies abutting fine working lands in a confounding juxtaposition. Many of the county's gullies developed in ways that appeared to be unrelated to human action or all out of proportion to any human actions that might have initiated them. Indeed, for many local residents the development of Providence Canyon and the rest of the county's caving gullies must have seemed the result of mysterious forces. Under the circumstances, it was not surprising or even completely disingenuous that locals, in their efforts to create a national park in the county, insisted that the "caves" were natural. They were not, of course, at least in any pure sense of that term. But the nature of the landscape, it soils and subsurface geology and hydrology, did have a lot to do with the gullies of Stewart County.

While I have found no evidence that a dripping barn roof caused the gullying that became Providence Canyon, there is nonetheless strong evidence that some of the county's gullies formed for reasons having little to do with agriculture's peculiar regional pathologies. Sam Singer, a prominent Stewart County resident, tipped me off to this when he shared with me a photograph (figure 25) that shows a gully headwall dropping directly from a crowned dirt road with drainage ditches on either side. The photograph indicates clearly that road drainage was the precipitant of at least one arm of Providence Canyon. Presumably the concentrated flow of water in the roadside ditches, or perhaps even in wagon wheel ruts on the road, wore down and broke through the Clayton clay layer and sent water into the Providence sands, initiating gullying. The *Stewart-Webster Journal* corroborated this dynamic more generally for the county in 1938, noting that the "canyons of quickest growth are those where footpaths or roadways caused the soil to be exposed to the washing effects of flowing water."[47]

Roads could be problematic in the rest of the South as well. H. Andrew Ireland, a geologist with the Soil Conservation Service who in the late 1930s conducted an investigation into the history of Georgia's other famous erosional landform, Lyell Gully, concluded that it was caused by concentrated water flowing in a roadside ditch. "Observations have shown," Ireland noted, "that

Figure 25. Gully development as a result of road drainage. This photograph suggests that concentrated water flow in roadside ditches contributed to gullying in Stewart County. Photograph courtesy of Sam Singer.

roads and road ditches are among the chief causes of gully-cutting because of the concentration of run-off and the lack of vegetation in them."[48] Arthur R. Hall also commented that the "most conspicuous culturally-induced erosion features of the Piedmont area are the gullies caused by old roads."[49] Hall even indirectly addressed another of the dubious origin myths for Providence Canyon: the possibility that Indian trails might have caused gullying. While he noted that he had "found no contemporary allusions . . . to gullying caused by Indian trails" in the South, he thought it "reasonable" that "there might have been places along trails where gullies could develop."[50]

Recognizing that roads and trails could be culprits in creating deep gullies does not absolve farmers from responsibility for them. In fact, we need to recognize that the gullying caused by roads and road drainage may have been exacerbated by the new hydrological cycles and increased runoff that came with land clearance for agriculture. But it does call into question whether it is accurate to use such places to visually represent the environmental impacts of southern plantation agriculture. Again, that does not mean that plantation agriculture in the South did not cause substantial erosion, because it did. My point is that the most spectacular gullies throughout the South turn out to be tricky markers of poor agricultural practices precisely because so many of them were caused either by roads and road drainage, or by conservation works such as plowed contours, ditches, and terraces.

Local geological conditions rather than regional generalities may better explain other deeply gullied parts of the South as well. The Piedmont, for instance, was also riddled with pockets where the subsoils were essentially structureless. Known as "sugar soils" because they seemed to melt under heavy rainfall, they eroded much more quickly and deeply when the forces of erosion reached them. Such isolated areas of caving gullies had as much to do with these structureless subsoils as they did with erosive land use. Not surprisingly, Greene County, Georgia, and Fairfield and Union Counties in South Carolina are places where such conditions occurred, places that, as we have seen, became well known for their large gullies.[51] The extreme gully erosion in the Loess Hills of Mississippi and Tennessee is also best explained with a localized analysis. This landscape is composed of Midwestern prairie loess that had been wind-deposited onto the hills adjoining the Mississippi Delta. These soils were fertile and well drained, which made them attractive to cotton farmers, but they washed easily. Moreover, they were sometimes underlain by unconsolidated sand, which heightened their propensity for gullying.[52] While sheet erosion and moderate gully erosion seem to have been widespread throughout the plantation South, then, extreme gullying mostly occurred in areas with unusual soil and underlying geological profiles. Again, my aim in making this point is not to absolve humans of responsibility from creating these spectacular gullies, for in every case human actions triggered them and humans could have prevented them. Rather, it is to point out that these gullies occurred in areas that were primed for such erosion, and that the human triggers were not always what we would call egregious land use practices.

In making sense of the spectacular gullies of Stewart County and other pockets of extreme southern gullying, it might be useful to abandon regional generalizing for a moment and examine areas, outside of the South, where human land use collided with similarly unstable soils and subsoils. One of the best examples within the United States is the Driftless Area of Wisconsin (and several adjoining states), another loess region that was badly dissected by gullying as a result of settler agriculture. In fact, the Coon Valley, a part of the Driftless Area, became the nation's first large-scale watershed conservation project—the Coon Creek Demonstration Area—under the auspices of the Soil Erosion Service in 1933. Human land use clearly set in motion the disastrous gullying that occurred in the area, although in this case it was midwest-

Figure 26. Gully erosion in the Driftless Area of Wisconsin. This 1937 photograph shows severe gully erosion near the town of Melrose, Wisconsin. Photo courtesy of the Natural Resources Conservation Service, USDA.

erners, many of them careful immigrant farmers, who caused the problem. They certainly were guilty, to borrow the evocative phrase of historian Neil Maher, of "square farming on round land," of an agriculture that did not take into account the rough topography and steep slopes of the Driftless Area. But the magnitude of the region's erosion also had a lot to do with its unusual topography and geology. The Driftless Area escaped the last glaciation, and so, unlike other parts of the upper Midwest, its topography was not ground down and smoothed out by the action of glaciers. As importantly, many of the high terraces of the region had loess soils underlain by highly erodible sand, not unlike the Loess Hills of Mississippi and Tennessee. Unsurprisingly, many of the worst gullies developed in those particular areas, whose soils and subsoils also disappeared like "melting sugar." The results were dramatic "arroyo-like channels" (see figure 26).[53] A larger lesson that Providence Canyon might communicate, then, is that the worst gullies not just in the South but throughout the

United States occurred in areas where human land use encountered landforms that were geologically and topographically unstable. Conditions in the South, from soils and climate to land use, may have made that region more susceptible to gullying than others, but the South was not the only place that experienced such dramatic soil loss.

Two further examples suggest the dangers of misreading gully causation and moralizing in facile ways about gullies as symbols of land abuse. The first comes from the region that, along with the South, most concerned American soil conservationists prior to the development of the Dust Bowl in the early 1930s: the American Southwest and particularly the lands of the Navajo Reservation. By the late 1920s and early 1930s, conservationists had grown alarmed that the Navajo range was in decline as a result of overgrazing. In June 1933, Franklin Roosevelt's new commissioner of Indian affairs, John Collier, sent Hugh Hammond Bennett to conduct a whirlwind tour of the reservation, and Bennett confirmed that the erosion there was serious. He not only found ubiquitous signs of sheet erosion but also dramatic arroyo cutting that had transformed formerly lush lands into a desert dissected by deep gullies. The solution that he and other federal officials proposed was a substantial reduction of Navajo sheep herds, under the logic that erosion and range decline were the clear results of overgrazing. There was some truth to the conservationists' diagnosis, but they also were too quick to assume that grazing had destroyed an Edenic landscape and that all the erosion they witnessed was the product of overgrazing. The Navajos challenged the diagnosis, insisting that the erosion conservationists worried about was the result of drought, not overgrazing. There was science to support their belief. An intense drought in the late nineteenth and early twentieth centuries followed by a novel pattern of high-energy storms proved particularly erosive in a landscape with diminished vegetation and with soils and subsoils that were particularly erodible. Today scholars largely agree that while grazing did affect erosion rates on Navajo rangelands, it was climate change that was critical to initiating the arroyo cutting that transformed much of the region. But New Deal conservationists assumed that only human agents could be to blame for such profound change, and so they went ahead with stock reductions that proved traumatic for the Navajo people, reducing them to economic dependency while doing little to improve range conditions. The result was an inaccurate condemnation of land use practices and an unjust and largely ineffective conservation intervention.[54] While the social dimensions of the Navajo case were entirely different from what occurred in Stewart County, it nonetheless provides a cautionary tale about the importance of explaining erosion in both cultural and environmental terms. Characteristically, the New Dealers saw poor human land use at

Figure 27. Gully erosion on the Navajo Reservation. From *Navajo Annual Extension Report*, 1936. Courtesy of the Records of the Bureau of Indian Affairs, National Archives.

work. For their part the Navajo blamed the climate and saw range decline as natural. The truth, as at Providence Canyon, lay somewhere in between, in the complex interactions of humans and environmental forces.

An equally compelling case of conservationists misreading the causes of gullying comes from the landlocked southern African nation of Lesotho, where massive gullies, known as "dongas," began to appear during the late nineteenth century and then spread alarmingly during the first half of the twentieth century. By then, British colonial officials became so concerned about the expanding network of gullies, which they took to be the products of the "primitive" farming techniques of the Basuto people, that they mounted a program to get local farmers to build engineered conservation works. But, in doing so, colonial conservationists missed several crucial facts. First, while the Basutos were farming in ways that stressed the land, such practices resulted not from their allegedly primitive ways but from the various pressures put upon them by colonial expansion, new crops and technologies, land competition, market intensification, off-farm labor demands, and even military conflict. Second, at the beginning of the twentieth century only about 10 percent of the landscape had been affected by incipient gullying, and many of the gullies appear to have been caused by the region's proliferating network of poorly constructed roads and trails, not by African farmers. Finally, and most tragically, colonial conservation ended up substantially *increasing* donga development. Officials insisted on the construction of engineering works such as terraces and diversion furrows, which transformed Basuto agricultural lands to solve a problem that did not yet exist for many of these lands. These engineering works resulted, in the words of historian Kate Showers, in "the collection and concentration of large amounts of previously unchanneled water." Colonial conservation interventions largely caused the gullying they were designed to prevent. Here again is a telling example of how conservation officials, set on a narrative of negligent human actors degrading otherwise productive lands, could misread landscapes of spectacular erosion and even exacerbate them through conservation works. The case of Stewart County is quite different, but there are important parallels between the two cases, not the least being the similar roles played by roads and conservation works in producing deep gullies.[55]

As much as it was at the center of a region whose agriculture was destructive of soils, then, Stewart County also was a unique place, one node in a larger world of extreme gullying whose historical lessons should lead us past stark nature-culture moralizing. By thinking locally about Providence Canyon's history, we can usefully interrogate the narrative of southern regional distinctiveness that many, myself included, have applied to it. And by thinking in an extraregional comparative way, we can reaffirm the importance of the local in explaining extreme erosion events. The distinction that Stuart Chase made between natural and human-made erosion, the distinction that seemed to disqualify Providence Canyon from park status, becomes muddled in this case. Providence Canyon's erosion was, in some senses, both geological and human induced.

The simplest lesson that emerges from this journey through Providence Canyon State Park's many possible interpretations is that the irony with which I began this study—that today an environmental disaster is protected as a park—is entirely reliant upon a rigid nature-culture dichotomy for its power. That irony, an inheritance of the 1930s, was rooted in and embodied by the twin efforts, by local boosters on the one hand and New Deal reformers on the other, to instrumentally objectify and simplify a site whose history is far more complex. In the case of the local boosters, this indictment is obvious: they attempted to sweep under the rug the role that human land use played in gully formation and to insist that Providence Canyon, like the nation's other great parks, was a product of nature. The New Dealers, for their part, were certainly right that human land use ignited these spectacular erosional processes and that staple crop farming had done a number on the soils of the entire region, but they were too quick to see Providence Canyon as only a cultural artifact, and as simply representative of a regional political economy out of whack. In such a rendering, the environment appears as little more than a passive and one-dimensional platform, when in fact environmental forces contributed to making the place so photogenic. To see such forces implicated in the creation of Providence Canyon is not to excuse or glide over human agency. It is to insist that human agency never exists outside of or divorced from its environmental context. Not only must southern environmental historians assiduously avoid the crude environmental determinism of an earlier historiography. We also must carefully revise a New Deal narrative, usually well illustrated, that continues to cast a long shadow over the region's environmental history.

Ultimately, my interpretation of what Providence Canyon should mean as a conservation area exists somewhere between the Grand Canyon and a sickening void, and it eschews the irony that flows from pairing these two simplistic historical visions. To embrace this postironic interpretation of Providence Canyon, we need to do for it something akin to what Walker Evans and James Agee did for the white sharecroppers of Hale County, Alabama: reject the facile moralism of the documentary impulse by simultaneously descending to a local level of detail and critically assessing the assumptions and narratives we bring to the place. However desirable it might be to make a spectacle like Providence Canyon stand for soil abuse throughout the South, such a representational strategy does not do justice to local environmental realities. As a park, Providence Canyon's most fruitful interpretive possibilities lie precisely in those zones where the natural and cultural meet and intermix. Providence Canyon's history warns against the interpretive simplification that comes with regional generalization, and it urges us to pay greater attention to the diverse and highly localized environmental histories within the larger region. As it turns out, Providence Canyon is far from alone among the South's conservation areas in having a more complicated environmental history than meets the eye.

EPILOGUE

The Ecology of Erasure

THE ALCOVY CONSERVATION CENTER, which is due east of Atlanta and just to the northeast of Covington, Georgia, is the headquarters of the Georgia Wildlife Federation (GWF), one of the oldest and most influential conservation and environmental education organizations in the state. Founded in 1936, largely as a sportsmen's group, GWF has a long and accomplished history of advocating for the preservation of Georgia's most important natural areas. The Alcovy Conservation Center, and what one GWF employee has described as "the 115 acres of pristine forest, meadow, and river swamp" that surrounds it, is one of those areas.[1] The environmental showpiece of the GFW property is a tupelo gum swamp at the confluence of the Alcovy River and Cornish Creek. A tupelo swamp is a rare thing to find on the Piedmont, let alone in an area on the verge of being swallowed up by Atlanta's eastward sprawl. I first visited on a November day, and even in its dry and leafless autumn state I found the gum swamp to be a magical place (figure 28). So did the many others who argued for its preservation and for the preservation of thousands of additional acres of swamplands along the Alcovy River corridor during the late 1960s and early 1970s.

The fight to save the Alcovy River swamps was a classic postwar environmental battle. In the 1960s the Soil Conservation Service, under the authority of the benign sounding Federal Watershed Protection and Flood Prevention Act of 1954, embarked on a program of channelizing small rivers and streams in the name of "stream improvement." These efforts were part of the SCS's Small Watershed Program, and its channelization activity accelerated in the early 1960s when the Kennedy administration directed more funding its way. The ostensible purpose of these channelization projects was to drain and prevent the flooding of valuable agricultural lands by straightening, widening, and clearing obstructions from rivers and streams. Channelization also involved draining swampy riparian areas and clearing banks of vegetation, often as far back as one hundred feet. Stream channelization, which represented a significant postwar expansion of the SCS's soil conservation mission, turned meandering watercourses into linear engineered ditches.[2]

Like so many federal programs launched in the name of conservation during the immediate postwar years, stream channelization was misguided. From

Figure 28. Tupelo Swamp, Alcovy Conservation Center. Photo by author.

an economic standpoint, the program expended considerable taxpayer money to bring small amounts of land into productive agricultural use while the federal government was simultaneously paying other farmers to take land out of production. Indeed, the SCS Small Watershed Program was getting up to speed just as Congress created the federal Soil Bank Program in 1956 (an important predecessor of today's Conservation Reserve Program), which paid farmers to retire unneeded, and often badly eroded, croplands and plant them in trees for soil conservation purposes. The Soil Bank Program, as we will see, had an outsized impact on the South. SCS stream channelization, then, benefited a few farmers with publically funded projects that worked at cross-purposes with other federal acreage-reduction programs aimed, in part, at curtailing agricultural production. In fact, some of the farmers who stood to benefit from stream channelization were themselves already receiving federal payments to keep other lands out of production. While channelization produced little, if any, net economic benefit, it caused considerable ecological damage to riparian systems. This was particularly true for the Southeast, which played host to a remarkable 70 percent of these SCS channelization projects. As wetlands historian Ann Vileisis grimly concludes, "The Small Watershed Program converted wildlife habitat into unneeded cropland and barren riparian zones—all at a costly burden to the taxpayers."[3]

As a result of the habitat destruction both caused and threatened by these ill-conceived projects, many southeastern state fish and game agencies, private conservation organizations such as the GWF, and concerned citizens organized in protest. In the process, state fish and game officials and their allies crafted a series of innovative arguments for the preservation of watery ecosystems that not only encouraged many to reconsider the ecological values of wetlands, long derided as useless environments in need of improvement, but that also argued for what ecologists and policy makers would today call the "ecosystem services" that such systems provide. It is no exaggeration to say that SCS stream channelization projects helped to produce an environmentalist sensibility in the Southeast, just as other postwar efforts by federal environmental management agencies—the extensive timber cutting that the Forest Service facilitated on the western national forests, the large-scale dam building projects proposed and completed by the Army Corps of Engineers and the Bureau of Reclamation, and the chemical pest control methods undertaken and encouraged by the U.S. Department of Agriculture—created storms of environmental protest in other parts of the country (and sometimes in the South as well).[4]

One of the projects that the SCS proposed was the channelization of the Alcovy River. At the request of area farmers, the SCS sought to "improve" more than seventy miles of the Alcovy and its tributaries, including Cornish Creek, at a cost of several million dollars. Sensing that this project was both economically unsound and ecologically destructive, a group of state officials and private citizens came together to oppose the plan.[5] This portion of the Alcovy River was already under consideration for designation as a State Scenic River, and environmental activists were incredulous that the SCS—a conservation agency, after all—failed to see any value in protecting the river corridor. Activists highlighted the Alcovy's threatened swamps and their many recreational, ecological, and economic benefits. Their goal, simply put, was to convince others that the river produced more value left as it was than it would once the SCS channelized it. Their advocacy coalesced in 1970, the year of the first Earth Day, when environmental concerns took the nation by storm, and it came in opposition to a federal conservation agency that had been at the center of New Deal environmental reform. In many ways, the fight to preserve the Alcovy River swamps evidenced how much environmental politics in the United States had changed since the 1930s.[6]

The great champion of the Alcovy cause was Charles H. Wharton, an ecologist at Georgia State University and one of Georgia's most important conservation activists at the dawn of the environmental era. Wharton wrote passionately about the value of river swamps, particularly in his 1970 tract *The Southern River Swamp: A Multiple Use Environment*, which came as a direct response to the Alcovy channelization proposal. *The Southern River Swamp* presented a remarkably broad and multifaceted case for appreciating the values of such places. "The river swamps are ideal examples of what we mean by 'open space,' 'green belts,' and 'natural corridors,'" Wharton instructed. "They may function in many ways: sponges for regulation of the vital water cycle, giant kidneys for waste purification, convalescent wards for the esthetically ill, outdoor classrooms for school children and oxygen machines for air quality." Wharton also lauded the biological productivity of river swamps. They supported abundant fish and game populations, harbored endemic and sometimes rare species, and provided vital corridors for species migration. He compared their productivity to salt marshes, another of Georgia's wetland environments then being celebrated for their ecological values by figures such as Eugene Odum, the pathbreaking University of Georgia ecologist (and son of the famed regionalist Howard Odum), and John and Mildred Teal, whose now-classic book *Life and Death of a Salt Marsh* had appeared the year before. Finally, Wharton insisted—just six years after the passage of the Wilderness Act and two years after the National Wild and Scenic Rivers Act—that the "river swamp is the last environment in the Southeast providing an accessible wilderness experience for a large part of our urban and rural population."[7] Wharton's was a wide-ranging and convincing case for leaving the region's river swamps as they were, and critical to the logic of his advocacy was his insistence on their deep-temporal naturalness. "Most swamps," he speculated, "are probably thousands of years old."[8]

Not everyone agreed with Wharton's conviction here, at least when it came to the Alcovy swamps. Stanley Trimble, who had completed his master's thesis in geography at the University of Georgia in 1969 and was beginning his influential dissertation on Piedmont soil erosion during the plantation era, offered a different interpretation of the swamps. Trimble's thesis had focused on the Oconee River just to the east of the Alcovy watershed, and he had found that most of the swampy lands there had resulted from culturally accelerated sedimentation produced by upland erosion in the nineteenth and early twentieth centuries.[9] Trimble suspected that the Alcovy swamps had similar origins, and so he embarked on a historical and geographical study, whose results would also be published in 1970.[10]

To determine the age of the Alcovy's swamps, Trimble mostly relied on historical sources. He began with the original survey plats produced by the office of the surveyor general of Georgia, which depicted the presettlement landscape of the region. The surveyors who created those plats had been under pressure to produce accurate results and particularly to depict lands that would not be suitable for agricultural settlement. In fact, they could be held liable for their mistakes, so Trimble figured he could trust their work. While these surveys and the accompanying field notes documented a few swampy areas along the Alcovy River, they were insignificant compared with the extent of the river swamps by the 1970s. Moreover, there was no sign of a swamp at the confluence of the Alcovy River and Cornish Creek, where the GWF's prized gum swamp sits today. Trimble also looked at land sale records. He discovered that the Alcovy's bottomlands sold at a premium and that early settlers often chose these bottomlands as home sites, more evidence that they were not swampy in the early nineteenth century. He then examined Civil War records, and he found no evidence of a swamp at an established ford across the Alcovy where General Sherman and his troops crossed in 1864, just to the south of where the Alcovy Conservation Center sits today, even though there was a swamp a half a mile wide at the same spot by the time of Trimble's study. Trimble's examination of early historical records raised serious questions about whether the Alcovy swamps, which environmentalists had come to prize, had even existed prior to the late nineteenth century.[11]

The historical records that Trimble consulted for the years after the Civil War betrayed the growing effects of sedimentation. An 1882 waterpower survey did not mention any swamps along the Alcovy, though it noted a sluggish river, a likely sign of stream aggradation from upland erosion. In 1891, a USGS topographical map of the area indicated swampy conditions developing in the lower area of the watershed, and in 1901 a soil survey of the Covington area noted "meadow land" subject to flooding along the Alcovy, though much of the river's bottomland was still being cultivated. By 1917, when the Geological Survey of Gerogia conducted the second of its two agricultural drainage surveys during that decade, they noted considerable swamping along the Alcovy in Walton County, upstream from the Alcovy Conservation Center's current location, and drainage records for the watershed clearly indicated that bottomlands were *becoming* swampy and unsuitable for cultivation. In fact, local drainage districts, organized after the 1911 state law that facilitated their creation, had cut ditches in the area to try to reclaim some of these bottomlands for farming, just as the SCS, at the request of local conservation districts, would attempt to do half a century later. The 1917 report specifically noted that the Cornish Creek Drainage District in Walton and Newton Counties had reclaimed 1,402 acres, and that the Alcova [*sic*] Drainage District upstream in Gwinnett and Walton Counties was in the process of draining 1,873 acres. These were two of the many Piedmont drainage districts formed during this period to try to drain bottomlands swamped as a result of upland erosion. By the mid-1930s, a USDA erosion survey made it clear that sedimentation had transformed the Alcovy: "Practically all of the bottom land along the Alcovy River has been covered by sand, or the channel has become so clogged with sand that the bottom lands are now swamps," the surveyor, D. H. Montgomery, wrote. Montgomery estimated that the upland areas around the Alcovy had lost an average of six inches of soil during its century of agricultural use, a figure that was consistent with other Piedmont estimates.[12]

With this historical evidence, and extrapolating from geographical studies of nearby watersheds like the Oconee, Trimble concluded that, despite their apparent naturalness, the Alcovy swamps were almost certainly of recent anthropogenic origin. "The flora and fauna of these swamps," he observed, "have developed to the point that the swamps are now mistakenly considered to be 'natural' and to be thousands of years old."[13] Trimble did leave an opening for Wharton and others who believed that the swamps were primeval: he noted that the best way to determine the age and depth of historical sedimentation along the Alcovy and its tributaries would be "to employ Carbon 14 or pollen dating methods, which are difficult and expensive processes."[14] A few years later, Wharton and a couple of Georgia State collaborators, hoping to prove that the swamps were indeed "natural," took six sediment samples and had them carbon-dated. The results supported their theory—the sediment dates ranged from 1785, just before settler agriculture came to the Georgia Piedmont, to as old as 6750 BC. But Wharton and his colleagues took so few samples, and there was so little method to their sampling, that Trimble and others effectively questioned their accuracy.[15]

Ultimately, the SCS backed away from their Alcovy channelization efforts, pulling federal funding for the project in early 1974.[16] The Alcovy Conservation Center's tupelo gum swamp and many of the Alcovy River's other swamplands were spared the worst effects of channelization, and they have become treasured refuges for scientists, educators, and wildlife enthusiasts. But their provenance remains unresolved. Wharton's latter-day supporters—such as the long-time GWF president Jerry McCollum, who graciously welcomed me to the conservation center on that November day—mostly offer ecological arguments to support their case for deep-temporal origins. Since tupelo swamps are usually found only on the Coastal Plain, and since some of the Alcovy swamps harbor endemic Coastal Plain species such as red buckeyes, bird-voiced tree frogs, and mole salamanders, some argue that these swamps must be habitat relics left over from a time when higher seas lapped at the fall line. Others, following Trimble and his considerable historical evidence, have continued to insist that these swamps are artifacts of anthropogenic erosion and that Coastal Plain species colonized them over the last century or so. In the late 1990s, at the end of his long and distinguished career, Eugene Odum tellingly opined of the Alcovy swamps: "Those fertile wetlands are the result of our upland folly. We don't know how old the trees are—maybe 100 years old. That's the time at the beginning of this century when we destroyed our soil." There is even speculation that Georgia's famed coastal marshes, now providers of critical ecosystem services, may themselves have been augmented by that very same upland folly.[17]

If we accept the logic of Trimble and Odum, then we might be tempted to conclude that the Alcovy River swamps qualify as an ironic conservation area on a par with Providence Canyon. But that would be a mistake. Again, irony is too limiting an interpretive approach for such an engagingly complex place. While it would be satisfying to know definitively the origins of these swamps, the more important project may be to critically reconsider what is at stake in the natural-unnatural distinction at the heart of this debate. While it was not Trimble's purpose to serve the SCS channelization effort, his argument that the Alcovy swamps were artifacts of anthropogenic erosion threatened to bolster the SCS cause. If these swamps were not "natural," how could they be worth protecting? For the proponents of preserving these swamps, on the other hand, their "naturalness" has been rhetorically powerful. In a moment of candor, Jerry McCollum, who was admirably open minded about the debate when we spoke, suggested that proclaiming the deep-temporal naturalness of these swamps served the GWF's purposes. He seemed reticent to entertain the Trimble thesis because he worried that it would weaken GWF's preservationist position. The fight over the Alcovy swamps is thus a powerful example of the ways in which environmental debates can be distorted by strict divisions between what is natural and what is not.

For me, the compelling lesson of the rare tupelo gum swamp at the Alcovy Conservation Center, and the debate about its provenance, is that we ought to move beyond the moral authority of a nature before history. As with Providence Canyon, the most interesting way to interpret the Alcovy swamps may be as artifacts of both nature and culture, of human and nonhuman forces intertwined and acting across time. Surely the GWF's tupelo swamp, whatever its origins, is in many ways a natural place. Nobody planted it or populated it with wildlife. The soils and hydrology may be artifacts of past human land use, but the forested habitat that exists there today is otherwise minimally influenced by human action or intent—except, perhaps, by enlightened restraint. Why does it matter, from a conservation standpoint at least, whether the tupelo gum swamp and the rest of the area's swamps are a hundred years old or thousands of years old? If they are havens for wildlife, safe harbors for rare species, providers of an array of ecosystem services, and recreational refuges, isn't that enough to argue for their protection? By raising these questions, I do not mean to suggest that time and human influence are unimportant variables in assessing the conservation values of particular landscapes. To the contrary, I believe that breaking down stark nature-culture divides will allow us to develop conservation criteria that deal with time and human influence in subtler and more satisfying ways.

One of the problems with the habit of valuing only a nature before history is that such an approach suggests that any human influence is degrading. If Trimble is right, preservationists might have to forsake the powerful narrative that the GWF's tupelo swamp is a place of pure nature set apart from history. They may have to give up the easy normative linkage between the "is" of deep-temporal naturalness and the "ought" of preservationist advocacy. But there is much to be gained in seeing the swamp as a valuable though novel ecosystem influenced by, though not wholly derivative of, human land use. Here is another case of a beautiful place that may have come to us, phoenix-like, as a result of processes of environmental degradation. One might even read today's Alcovy River landscape in the same way that the poet Thomas Jefferson Flanagan interpreted Providence Canyon—as a hopeful monument that embodies both tragedy and beauty. Acknowledging the likely history of anthropogenic erosion in the creation of the Alcovy swamps better connects their laudable protection to the less-than-laudable histories of the lands that surround them. Rather than creating parks and conservation areas that separate nature from history, here is a remarkable manifestation of how we might do so by bringing them together. Like Providence Canyon, the Alcovy River swamps deserve such an interpretation.

The Alcovy swamps embody another important historical development as well: the ways in which the long history of human-induced soil erosion in the South has faded from view since World War II. When soil conservationists first surveyed the region, they encountered a veritable "theater of environmental disasters," to borrow Albert Cowdrey's description of the South in the first few decades of the twentieth century.[18] But since then, it has gotten easier by the year *not* to see the history of erosion in the landscapes of the former plantation South, and thus to forget that history. As we rethink the usefulness of stark divisions between nature and culture, then, we also need to attend to what we might call the ecology of erasure: the ways in which environmental forces have, with much human encouragement, obscured the South's history of human-induced erosion from view.

Providence Canyon is a different sort of site than the Alcovy River swamps—

Figure 29. The ecology of erasure. One of the unusual sights on the loop trail around Providence Canyon is an old automobile dump that provides dramatic evidence of reforestation and the ongoing ecology of erasure. Photo by author.

not a seemingly natural area with more cultural history than meets the eye, but a seemingly cultural place with more natural history than meets the eye. Indeed, a big part of its interpretive appeal has been its continuing visibility in a landscape that has swallowed up many signs of past erosive land use. It is a place where we can remember what the Alcovy swamps encourage us to forget. But as agriculture has migrated away from Stewart County and more of the county's land has reverted to forest, even Providence Canyon is threatened with environmental as well as cultural obscurity. These processes of erasure and forgetting raise a management question that has troubled me as I have become attached to the interpretive possibilities of Providence Canyon. How far ought we to allow environmental forces to proceed in reclaiming a preserved Providence Canyon if the result will be a diminishment of its power as a site of historical interpretation? Should managers maintain Providence Canyon as a stark visual spectacle, or should they allow reforestation and soil aggradation to slowly heal the site, obscuring its history and blunting its lessons? While I have tried to wed natural and human history in my interpretation of Providence Canyon, there are also inherent tensions between natural and historical preservation, between environmental processes that are having their way with the site and fading memories of its more human past.[19]

These tensions between nature and culture, between preservation and restoration, between ecology and memory lead me to a final set of lessons that Providence Canyon might teach us about the coming of conservation to the South and its relationship to what the eminent southern historian Thomas D. Clark called the "greening of the South." Providence Canyon's preservation, while not quite the fulfillment of the 1930s booster visions, was of a piece with a larger trend across the region. "For broad sweeps of the South," Clark concluded, "the ancient red-hill badge of shame, so harshly condemned only a half century ago by Big Hugh Bennett, has now been tucked gracefully beneath a blessed tree cover."[20] Over the last century, the South, like the Northeast and the upper Midwest, has seen an "explosion of green," as forests have returned to and reshaped huge swaths of formerly agricultural land.[21] The South's state and national parks and forests today are mostly the products of these processes, and almost all of them have histories of human land use hidden beneath their mantle of trees. This is a story that has been well told for the region's Appalachian conservation areas, but it has been less well recognized for much of the rest of the region. And while Providence Canyon may be unique in its preservation of soil erosion as scenery, a surprising number of today's southern landscapes of conservation exist where once there was substantial human-induced soil erosion. In its correlation between past destructive land use and current conservation protection, Providence Canyon is far from alone.

To understand this correlation, we need to take a quick detour into the history of another of the South's environmental traumas. From the 1880s to the 1920s, even as southern farmers produced the region's most destructive pulse of erosion, the South experienced rapid industrial deforestation. Beginning in the late 1870s, as the Great Lakes forests gave out and as repeal of the Southern Homestead Act made the cheap purchase of unlimited public domain lands possible, the timber industry came to the South with a vengeance, with timber companies cutting more than 150 million acres of forest on the margins of the plantation crescent. The timber industry cut extensively across the longleaf pine belt of the Coastal Plain, which had been dominated by the naval stores industry and open range grazing but had seen only spotty agricultural development. They moved aggressively into Appalachia, exploiting its vast hardwood reserves. And they cut out the region's remaining bottomland hardwood forests, whose wet intractability had kept them safe from agricultural transformation. As cotton was digging its nails deeper into the plantation crescent's thinning soils and hanging on for dear life, its forested fringes came crashing down. By the 1920s, the South's forestlands were in as sorry shape as its agricultural lands. This terrible conjuncture precipitated the movement of federal conservation into the region.[22]

The South's land crisis, and the various conservation responses to it, differed by subregion. The industrial cutting of forests in southern Appalachia in the late nineteenth and early twentieth centuries—and the questionable presumptions by land planning experts that Appalachian agriculture was both environmentally destructive and economically marginal—heightened conservationist desires to create national forests and parks in the South's mountainous regions, even if that meant displacing long-time residents. In 1911, as soil surveyors were beginning to piece together their portrait of the eroded South, Congress passed the Weeks Act, which allowed the federal government to purchase eastern lands for the creation of new national forests. Under the Weeks Act, the federal government purchased large areas of mostly cutover land in the mountainous South, creating new national forests and returning those lands to forest over the course of the twentieth century. By the late 1920s and 1930s, there were movements to create the Shenandoah and Great Smoky Mountain National Parks, again mostly on lands that had been previously farmed or cut over during the industrial timber era. The southern Appalachian landscapes of conservation that emerged in the early twentieth century grew from the ashes of environmental transformation, and like Providence Canyon, they often contained deep human histories and enduring land use legacies.[23]

As industrial timber cutting waned in the 1920s and 1930s, there was a different conversation about what to do with the cutover pinelands of the Coastal Plain. Longleaf pine woodlands had once covered an estimated seventy to ninety million acres of the Coastal Plain, constituting one of the most conspicuous of North America's biomes. These longleaf woodlands had long been fire maintained, but the desires of the timber industry and foresters to quickly reforest these lands led to efforts to evict fire from the landscape. While there were some moves to create national forests on cutover portions of the Coastal Plain, conservation more often came to this part of the South in the guise of fire suppression. Longleaf pines need fire to regenerate, as does the biodiverse groundcover that is one of the hallmarks of this forest type, but the foresters' favored replacement species, slash and loblolly pines, do not tolerate fire well, particularly at the seedling stage. Forestry interests favored slash and loblolly because these species were excellent colonizers of disturbed and degraded land, quickly growing up in thick stands even in the absence of replanting efforts. Moreover, and unlike longleaf at the time, these other pines could be planted with success and grown in close quarters, and they grew to maturity quickly in the favorable southern climate. The generative powers of these old-field pines quickly transformed lamentations about southern forest destruction into booster visions of a forested future for the region. While southern agriculture did migrate to and intensify within portions of the Coastal Plain during the postwar years, plantation pine production took over much of the rest of the region. Indeed, in another remarkable conjuncture, a serendipitous new industry, the pulp and paper industry, came to covet this "second forest." Pulp and paper proponents pushed hard for the region to embrace pine trees as its next economic panacea. But these pine monocultures, a major contributor to Clark's "greening of the South," were a far cry from the longleaf pine woodlands they replaced, both ecologically and aesthetically.[24]

One of the biggest boosters of this transformation was none other than Jonathan Daniels. In *A Southerner Discovers the South*, Daniels had departed from his memorable encounter with Providence Canyon and traveled to the Savannah laboratory of Dr. Charles Herty. Herty was a native Georgian (born, Daniels wrote, "in Milledgeville, Georgia, where Lyell the geologist saw the big ditch") who is today best known for founding the Pulp and Paper Laboratory in Savannah and for fathering the pulp and paper industry in the South. The region already had a kraft paper industry, but Herty dreamed of challenging the dominant position that the Northeast and Canada enjoyed in the production of white paper for newsprint and other uses. During the early 1930s, Herty's laboratory successfully solved the technical problems associated with transforming high-resin southern pines into white paper, and Herty himself became a tireless booster of the pulp and paper industry as the key to realizing the long-deferred dream of a New South. This was a development of great interest to a southern newspaperman like Jonathan Daniels.

For the pulp and paper industry to really take off in the South, it would need a large and proximate supply of pulpable timber. The "second forest" regenerating all over the region inspired a concerted effort to grow trees for this insatiable industry.[25] In 1957, exactly twenty years after his visit to Herty's lab, Jonathan Daniels made a second journey through the South to discover how the pulp and paper industry was transforming the region. The result was a short book, commissioned and published by the International Paper Company, titled *The Forest Is the Future*. Daniels found that the industry had grown "almost as quickly as the loblolly pine." By the mid-1950s, forest products had a higher value in the region than agricultural crops, and those products had created a "big parade" of paper companies into the South. These companies bought up a considerable southern land base, more than twelve million acres according to Daniels, for short-rotation pine production. In fact, the industry pioneered many of the forestry techniques that rapidly turned so much of the Southeast into pine plantations. But Daniels also noted that the majority of forest production was occurring on private land, much of it former agricultural land converted to forestry.[26]

The pulp and paper industry was not the only force contributing to this "greening" trend. Just as the Weeks Act was critical to conserving the mountainous South, so other federal policies shaped the greening of the rest of the region. The Clarke-McNary Act of 1924 expanded on the Weeks Act by encouraging federal-state cooperation on fire prevention, providing aid to states and private landowners to encourage the reforestation of denuded lands, and expanding the purchasing powers of the Forest Service. Various New Deal policies furthered this greening trend. The Agricultural Adjustment Act (AAA) of 1933 paid farmers to remove land from production, largely to decrease crop surpluses and raise commodity prices. In the South the result was that landowners tended to take their least productive and most eroded lands out of production, and to put tenants off the land in the process, funneling government payments into more intensive farming methods on their best lands. Federal conservation programs such as the Civilian Conservation Corps and the Soil Conservation Service also made tree planting (and kudzu planting) one of their chief activities, much of it on private lands taken out of farming. When the Supreme Court ruled that the original AAA was unconstitutional, Congress reworked the program in the Soil Conservation and Domestic Allotment Act of 1936, which made soil conservation and erosion prevention the major constitutional rationale for paying farmers to remove submarginal lands from crop production. New Deal policies, then, were instrumental in propelling the region toward a more forested future.[27]

After World War II, as the pulp and paper industry increased its efforts to reforest the cutover parts of the region, the Soil Bank Act of 1956 provided even more powerful incentives for southern landowners to plant former farmland in pine trees, producing what William Boyd has called "a golden age of southern forest regeneration." Like New Deal programs, the Soil Bank Act aimed to reduce agricultural overproduction and to encourage conservation. Under the act, farmers who planted some or all of their land in trees were reimbursed for 80 percent of their planting costs. They also received annual government payments to offset the loss of crop production revenues for ten years. In effect, the Soil Bank, while ostensibly a conservation program, allowed southern landowners to shift from crop to timber production by paying for the costs of tree planting and providing landowners with a source of income while their trees grew toward commercial maturity. As a result of this program, millions of acres of southern farmland, much of it in the eroded plantation crescent, went into forestry.[28] Along with the ecological voluntarism of old-field pines and the enticements of the pulp and paper industry, these government conservation programs produced forests where staple crop farming had long dominated. Thus did much of the South's agricultural past come to be hidden from view.

As the South grew up in trees, the federal government created, often in cooperation with the states, a slew of new conservation areas after 1930. The region's new national forests from this period, many of them within the Piedmont and Coastal Plain rather than in the mountains, are a case in point. Some of them, such as the Francis Marion National Forest in South Carolina, were the result of the federal government buying up cutover timberlands from timber companies for reforestation, though the Francis Marion contains within its bounds the still-visible remnants of old rice embankments that predated the industrial forestry era. But a more intriguing suite of national forests developed between 1930 and 1960 across the former cotton kingdom. These national forests came into being not to protect mountainous watersheds or to reforest lands in nonfarm regions that the timber industry had cut over. Rather, they sprang up precisely where human-induced soil erosion had been most acute. They are grave markers of a sort, memorials to a land-killing past.

The Sumter National Forest in upcountry South Carolina is a prime example of this phenomenon, for it contains within its bounds some of the most eroded and gullied lands of that state's cotton belt, including portions of both Fairfield County, which Hugh Bennett used as one of his constant exemplars of southern land abuse, and Union County, whose erosion Bennett profiled in a 1931 article. Federal purchase of lands within two discrete areas of the South Carolina Piedmont began in 1934, during the early New Deal years, enabled by the purchase provisions of the Clarke-McNary Act. President Roosevelt then established the Sumter National Forest by proclamation on July 13, 1936, uniting these two units under a single jurisdiction. Most of the purchased land was marginal or abandoned cotton land. Some of it was classified as cutover land, though even those lands had an agricultural past, having been abandoned a generation or two earlier to old-field succession. Much of the land was badly gullied, and federal purchase aimed to take such land permanently out of agricultural production to prevent further damage.[29] In 1947, the Sumter became home to the Calhoun Experimental Forest, created as a site for studying "forest stabilization of eroding agricultural soils and for the management of secondary southern pine forests."[30] As one of the Calhoun's early publications noted, the site of the experimental forest "was chosen because it represented the poorest Piedmont conditions."[31] Since 1957, the site has hosted a long-term soil-ecosystem study, now known as the Calhoun Critical Zone Observatory. The Sumter National Forest, then, is not only a reforested conservation area built atop some of the worst eroded soils of the South. It is also

Figure 30. Forest ranger and gully erosion on the Sumter National Forest, early 1950s. Courtesy of the U.S. Forest Service, USDA.

host to an important research effort designed to study the ecological dynamics of that very transformation.

Like the Sumter, a number of other national forests in former cotton regions of the South came into being during the Depression and New Deal years, when such land could be had at rock-bottom prices and when the federal government finally gained the capacity to put such lands into permanent conservation. Even before the election of Franklin Roosevelt, the Forest Service was purchasing lands for what would become the Homochitto National Forest in southern Mississippi, an area that had once been a part of the Natchez District, whose loess soils became badly eroded and gullied. "The history of these lands," the National Forest Reservation Committee wrote in 1930 of a portion of this purchase unit, "is the unsuccessful attempt to develop a farming community under the handicap of broken surface, concentrated precipitation, and friable and easily eroded soils soon depleted of natural fertility by active and excessive erosion."[32] Likewise, the Holly Springs National Forest to the north, just outside of Oxford, Mississippi, overlies soils that had suffered from terrible erosion and gullying.[33] This was roughly the same area where Stafford Happ and his colleagues conducted their pioneering 1940 study on accelerated stream and valley sedimentation. Buying eroded agricultural lands in the South and turning them into national forests was one of the notable conservation activities of the New Deal state.

Even more telling of this dynamic are several other southern national forests that grew from New Deal land utilization projects (LUPs), ambitious efforts to reengineer rural economies and rural land use in areas where poverty, racial inequality, and environmental degradation walked hand in hand. These lands took a more circuitous route to national forest status. LUPs had their roots in the work of the National Resources Board, whose 1934 report recommended that the federal government purchase seventy-five million acres of submarginal farmland and put it into permanent conservation. That report included the pioneering Reconnaissance Erosion Survey of the United States, part of whose purpose it was to identify badly eroded lands for federal purchase. The resulting program was more modest than originally planned; it included some 250 LUPs involving the federal purchase of 11.3 million acres in forty-five states between 1933 and 1946. But it was nonetheless a far-reaching effort to retire degraded lands (or lands threatened with degradation) and to rationalize land use on the acres that remained in production. These LUPs varied widely in their aims, size, scope, and management, but all were premised on permanent forms of agricultural resettlement calibrated to the productive capabilities of particular soils, and all sought to take the most eroded or vulnerable lands out of crop production if possible. The northern plains saw the largest purchases under this program, but the South ranked second with about 20 percent of the projects' purchased acreage. In many parts of the South, conservation areas of one sort or another are all that remain of these ambitious land-planning schemes.[34]

America's smallest national forest is the Tuskegee National Forest, which sits just to the northeast of Tuskegee in Macon County, Alabama. The town is famous for Tuskegee University, formerly the Tuskegee Institute, a historically black university founded by Booker T. Washington in 1881. From the institute's inception, Tuskegee officials worked with African American farmers in the area to encourage landownership and improve agricultural practices. But by the 1930s, despite a rise in African American land ownership and Tuskegee's best efforts, a large part of the county's land base was badly eroded. These conditions and the presence of the Tuskegee Institute made the area an attractive one for a New Deal land utilization project focused on African American farmers. The Agricultural Adjustment Administration and Tuskegee Institute officials first proposed what became known as the Tuskegee Planned Land Use Demonstration (TPLUD) in 1934.[35] In portions of Macon County,

Figure 31. "Gully erosion by check dam. Macon County, Alabama. Tuskegee Project." This 1937 photograph by Arthur Rothstein shows characteristic Black Belt gullying in Macon County, Alabama, on lands that were part of the Tuskegee Project. Courtesy of the Farm Security Administration / Office of War Information Photograph Collection, Library of Congress.

the report noted, "soil erosion is extremely pronounced and much havoc has been wrought to the land. Deep gullies, furrowed hillsides and sand-bottomed valleys greet the eye from every angle."[36] Some of the gullies in the county were as much as sixty feet deep.[37] In 1935, the federal government began buying up some of the worst land in the county for the TPLUD, eventually amassing 10,358 acres, one-third of which came from African American landowners. The project, headquartered at the Tuskegee Institute, moved several hundred families off submarginal lands and resettled some of them on better farmland in the area. Two-thirds of the acres purchased were planted in trees, another thousand acres went into pasture, and the rest into other land uses. Erosion control was necessary over the entire area (see figure 31). The TPLUD aimed to rehabilitate badly eroded land as well as the region's impoverished farm population. The former goal would be achieved with greater success than the latter.[38]

Despite an initial flurry of activity and enthusiasm, the TPLUD entered a period of sustained crisis by the late 1930s, as funding declined and federal priorities shifted. The TPLUD moved to the Soil Conservation Service in 1938, and for the next ten years the SCS assumed project management, although in practice officials at Tuskegee did much of the work. The Tuskegee Institute made several efforts to gain control of the project, but in the end they lacked the resources to do so. In 1949, the SCS leased the TPLUD lands to the East Alabama Soil Conservation District for a ten-year period. Then, in 1954, an act of Congress transferred the Tuskegee LUP lands from the SCS to the U.S. Forest Service. By this time approximately nine thousand acres of the land had grown up in forest. In November 1959 President Dwight Eisenhower created the Tuskegee National Forest. The existence of Tuskegee National Forest today attests to the success of the land rehabilitation goals of the TPLUD, but the social goals of the project were only partially achieved. While many families did successfully relocate to better farmlands, many others drifted away from the area. The Tuskegee National Forest thus stands as an ambiguous monument to the only New Deal–era LUP managed by African Americans and to the reforestation of badly eroded Macon County lands. In this case, the ecology of erasure has obscured the history of a fascinating social experiment.[39]

Georgia's Oconee National Forest, home to the Scull Shoals site and the Murder Creek watershed, also came into being as a result of the same 1959 proclamation by President Eisenhower. The Oconee was pieced together from lands purchased by Georgia's two largest Piedmont LUPs: the Plantation Piedmont Project, which included lands in Putnam, Jasper, and Jones Counties just to the north of Macon, and the North Central Georgia Project, which focused on Greene County. As H. H. Wooten, an economist with the USDA and a student of these projects, noted decades ago, these Georgia Piedmont projects "were in the Briar Patch Country, made famous by Joel Chandler Harris" in his Uncle Remus stories, a landscape of sharecropping, shifting cultivation, soil degradation, land abandonment, old-field succession, and wildlife that thrived on such disturbance and early-successional regeneration. In 1934, W. A. Hartman of the USDA opened an office in Eatonton, Georgia, Harris's birthplace, with the aim of making large submarginal land purchases in the lower Piedmont cotton counties that eventually hosted these two LUPs. Greene County elites were apparently enthusiastic about the idea of such purchases, for they sent a request to their congressional representatives asking that they "use their influence in securing an allotment to establish *a national park* in Greene County." It's not clear if they planned to feature any of the county's impressive gullies in their park plan, but in all likelihood this invocation of national park status was a confused reference to an emerging policy of converting some of the submarginal lands purchased by the federal government into state parks and other sorts of recreation areas.[40]

These two LUPs developed in the region not only when cotton agriculture was in steep decline but also as landowners and tenants, unable to make a cotton crop, were turning to their timber resources for cash. "Forest resources had been heavily exploited since 1920," noted a government report, and "[e]rosion was almost as active in the woodlands as in the open fields." These lands, then, presented in microcosm the twin land crises that faced the region during the Depression. Not surprisingly, the region's residents had been leaving in large numbers since World War I.[41] One of the most fateful departures was a well-to-do Greene County planter named James G. Boswell, who left in 1921 for Southern California, where he, his brother, and then his son, J. G. Boswell II, built the nation's largest privately owned "farm," an agricultural empire of more than one hundred thousand acres in San Joaquin Valley that came to dominate that state's cotton production. It was a migration that in many ways symbolized the transformations that have marked American agriculture since the New Deal. As cotton moved west with folks like Boswell, and as many unlike Boswell departed from Greene County, there was a lot of land available for land planning and conservation.[42]

In Greene County, the North Central Georgia Project fed into a late New Deal initiative known as the Unified Farm Program, an effort that Arthur Raper documented in his book *Tenants of the Almighty*. Initiated in 1939, the Unified Farm Program was a demonstration project that used local planning committees to coordinate the work of the myriad government programs then reshaping the county.[43] One part of that effort was land planning. As Raper recalled, a land use planning committee quickly put together a map of the county that showed which areas should continue in cultivation and which ought to be kept permanently in pasture or forest. The federal government stepped in with funds to retire submarginal lands, and the CCC transferred one of its African American camps to Greene County to control erosion and plant trees. Aside from conservation, the Unified Farm Program also included public health and sanitary measures, construction of improved housing and farm infrastructure, efforts to diversify agricultural production, canning and other home economics programs, and new schools. It was a comprehensive program of rural improvement.[44] The main contribution of the North Central Georgia LUP to Greene County's Unified Farm Program was the purchase of submarginal lands. It concentrated its efforts in the badly eroded northwestern part of the county, land adjoining the Oconee River, where it purchased nearly thirty thousand acres. Much of the land had already grown up in pines, and some of it had even been cut over in recent years. Today that land is Greene County's portion of the Oconee National Forest, transferred to the Forest Service from the Soil Conservation Service in 1959. It is land whose extensive erosion inundated the Scull Shoals site with sediment and undid its nineteenth-century prosperity.[45]

The Oconee National Forest is the most prominent conservation legacy of Piedmont Georgia's LUPs, but it is not the only one. Adjoining the national forest in Jasper and Jones Counties is the thirty-five-thousand-acre Piedmont National Wildlife Refuge, established by President Roosevelt in 1939 from LUP lands. Its now-mature loblolly pines support a population of endangered red-cockaded woodpeckers. Other Piedmont LUP lands became the Hitchiti Experimental Forest, a facility of the U.S. Forest Service's Southern Research Station; the Rock Eagle 4-H Center, which is today administered by the University of Georgia; Georgia Experiment Station lands; and a small Jones County recreational area. A third Georgia Piedmont LUP, the Hard Labor Creek project, is now Hard Labor Creek State Park, which was developed by the CCC and which also has a past history of extreme erosion that can still be detected with a discerning eye.[46]

These examples from the Georgia Piedmont have analogues throughout the former plantation South, where national forests, state parks, state forests, national wildlife refuges, national recreation areas, and other landscapes of conservation emerged from LUPs, direct federal purchase, and various other New Deal and postwar programs. As these examples suggest, it has not been unusual in the South for an area with a history of extreme erosion to become a state park, like Providence Canyon, or a national forest, like Oconee National Forest, or to be protected by other conservation designations. Indeed, one might argue that this has been closer to the rule than the exception. The apparent irony of Providence Canyon fades even further when we appreciate this correlation between landscapes with histories of soil erosion and conservation in the South.

This southern story has analogues in other parts of the country. Not only do a subset of other national forests mark areas where agriculture was tried and failed, including a total of 1.5 million acres of former LUP lands, but many state parks and state forests are the products of the restoration of former agricultural lands and industrial timberlands. Perhaps the most prominent and novel conservation category created from former LUP lands is our system of national grasslands. In 1960, the USDA created nineteen national grasslands from twenty-two former LUPs on the Great Plains, a system totaling four million acres. Four of these national grasslands are in the heart of the former Dust

Bowl region of the southern plains. America's national grasslands represent a novel conservation domain that also came into being as a result of soil conservation interventions.[47] From this understanding of southern conservation history and its analogues elsewhere in the nation, we can appreciate a larger lesson: federal conservation efforts emerged in the half-century between 1890–1945 not just to save the last best places or provide expert resource stewardship, but also to figure out what to do with those parts of the American landscape that would never be suitable to traditional agriculture or that had been degraded by agriculture or other forms of human land use.

As a conservation area, then, Providence Canyon State Park is not a southern anomaly filled with irony. In many ways it is a representative place.

Several months after visiting the Alcovy River swamps, as I was finishing an initial draft of this book, I took another field trip to a place that several friends told me that I had to see. Malakoff Diggins State Historic Park, which is in California's Sierra foothills not far from the gentrified former mining town of Nevada City, is Providence Canyon's western doppelganger. In the decades after the California Gold Rush, Malakoff Diggins (a shortened rendering of "Malakoff Diggings") was the site of the nation's largest hydraulic gold-mining operation. Hydraulic mining, which began at the site in the 1850s and became highly capitalized under the North Bloomfield Mining and Gravel Company, involved focusing pressurized water cannons on the area's hillsides to loosen gravel and ore, which was then processed through huge sluices. Hydraulic mining produced tremendous wealth in California, as tobacco and cotton had in the South, but its environmental consequences were profound. Not only did hydraulic miners blast away entire hillsides, leaving eroded moonscapes in their wake, but the debris produced by hydraulic mining choked the region's rivers and caused frequent downstream floods, particularly during spring runoff. (The mercury that processors used to extract gold also poisoned these waterways.) As a result, a lot of local farmland and some of the downstream towns suffered frequent flooding and sediment inundation. Much of the lighter sediment made it all the way down into the Sacramento River and its delta, thwarting navigation, causing major floods, burying prime farmland, and destroying lucrative fisheries. In the late 1870s, the environmental effects of hydraulic mining became so acute that Sacramento Valley farmers organized the Anti-Debris Association to fight back against the industry. Then, in 1882, a Marysville farmer named Edwards Woodruff, who owned a 1,700-acre wheat farm on the Yuba River, sued the owners of the Malakoff Diggins for damages. In a landmark environmental decision in 1884, *Edwards Woodruff vs. North Bloomfield Mining and Gravel Company*, U.S. District Court Judge

Figure 32. "Malakoff Diggings, Nevada County, California." This photograph by Carleton Watkins shows the impacts of hydraulic mining at the Malakoff Diggings in California. Courtesy of the Bancroft Library, University of California, Berkeley.

Figure 33. Erosional landscape at Malakoff Diggins State Park. Photo by author.

Lorenzo Sawyer decided in favor of Woodruff and enjoined hydraulic miners from discharging debris into the region's streams and rivers. The decision struck a major blow against hydraulic mining in California. It also suggested the rising power of agricultural interests in the valley below.[48]

While the historical forces that produced Malakoff Diggins were quite different from those that produced Providence Canyon, it too has become a park that preserves the spectacular results of human-induced erosion. In 1965, just a few years before the creation of Providence Canyon State Park, local activists successfully lobbied the state to create Malakoff Diggins State Historic Park. In their interpretation, the California Department of Parks and Recreation does little to celebrate the aesthetic results of hydraulic gold mining at Malakoff Diggins, though the images they deploy suggest that many park visitors will find the landscape at least as compelling as the historic town site. Like Providence Canyon, Malakoff Diggins is a little-known and out-of-the-way park, one that has also been threatened with closure as the result of a state budget crisis. But it too deserves to be famous, both for the spectacle it presents and for the lessons it can communicate about another chapter of our erosive past.

After spending an afternoon at Malakoff Diggins, reveling in its weirdness, I drove a few hours south to pay my first visit to Yosemite National Park. It was a true pilgrimage to one of the most iconic landscapes of American conservation, one that I had been meaning to make for decades. I spent two nights in a canvas cottage in the densely packed Curry Village and the better part of a day hiking the entire Yosemite Valley Loop Trail. On the morning of my departure, I drove up to Glacier Point. It was early May, and Glacier Point Road had just unexpectedly opened for the season that day, several weeks ahead of schedule. The weather was warm and the winter snowpack unusually sparse, and as I stood looking down upon the Yosemite Valley more than three thousand feet below, I thought of the famous photograph of John Muir and Theodore Roosevelt, taken in about the same place in 1903. It was an image that helped to define the first great age of conservation in American history: two men, on a precipice, bringing together passionate advocacy for wild places and the power of the modern administrative state. But as I gazed down at the glacially carved grandeur of the Yosemite valley and admired the waterfalls plunging violently toward the valley floor, my mind kept drifting back to Malakoff Diggins, and then to the American South and Providence Canyon. As sublime as Yosemite was that day, I realized that parks like Malakoff Diggins and Providence Canyon, very different testaments to the sculpting power of water, had become more interesting to me as an environmental historian. Neither will ever see the kind of visitation that Yosemite does, but they are parks that deserve our attention. Malakoff Diggins is not nearly as grand as Yosemite, and, of course, Providence Canyon is hardly comparable to the Grand Canyon. But each has its aesthetic charms, and each in its own way can help us come to terms with the fullness of our environmental history. They are places worth carrying with us into a new age of conservation.

NOTES

A NOTE ON TERMINOLOGY

1. Charles Aiken, *The Cotton Plantation South since the Civil War* (Baltimore: Johns Hopkins University Press, 1998).

INTRODUCTION. The Great Cut across the Face of Nature

1. Georgia State Parks and Historic Sites website for Providence Canyon, http://www.gastateparks.org/info/providence/, accessed October 10, 2014. The second quote is from an interpretative panel at the park titled "The Canyon Begins to Form in the 1830s." The final quote from the park video is taken from Dan Chapman, "Providence Canyon Escapes, but 5 State Parks Partly Closed," *Atlanta Journal-Constitution*, May 22, 2007.
2. Ellis Gibbs Arnall, *The Shore Dimly Seen* (Philadelphia: J. B. Lippincott, 1946), 63.
3. Chapman, "Providence Canyon Escapes."
4. Mary Ann Anderson, "Georgia Has Its Own 'Seven Natural Wonders,'" *Miami Herald*, May 27, 2007.
5. "Providence Canyon Showcased on American Highways," Georgia Department State Parks and Historic Sites, http://gastateparks.org/item/144001, accessed November 6, 2014.
6. On the story of *Let Us Now Praise Famous Men*, see William Stott, *Documentary Expression and Thirties America* (New York: Oxford University Press, 1973), 261–314.
7. Macdonald quote in ibid., 262.
8. Ibid., 261–66; Walker Evans and James Agee, *Let Us Now Praise Famous Men: Three Tenant Families* (New York: Houghton Mifflin, 1941).
9. This was a format repeated by others during the era, such as Arthur Raper's *Tenants of the Almighty* (New York: Macmillan, 1943).
10. Donald Worster, *Dust Bowl: The Southern Plains in the 1930s* (New York: Oxford University Press, 1979); Sarah Phillips, *This Land, This Nation: Conservation, Rural America, and the New Deal* (New York: Cambridge University Press, 2007); Neil Maher, *Nature's New Deal: The Civilian Conservation Corps and the Roots of the American Environmental Movement* (New York: Oxford University Press, 2008); Randal S. Beeman and James A. Pritchard, *A Green and Permanent Land: Ecology and Agriculture in the Twentieth Century* (Lawrence: University Press of Kansas, 2001); Finis Dunaway, *Natural Visions: The Power of Images in American Environmental Reform* (Chicago: University of Chicago Press, 2005); Marsha Weisiger, *Dreaming of Sheep in Navajo Country* (Seattle: University of Washington Press, 2009); and Geoff Cunfer, *On the Great Plains: Agriculture and Environment* (College Station: Texas A&M University Press, 2005).
11. Dunaway, *Natural Visions*.
12. Mart Stewart, "If John Muir Had Been an Agrarian: American Environmental History West and South," *Environment and History* 11, no. 2 (2005): 139–62.

13. Ulrich Bonnell Phillips, *Life and Labor in the Old South* (Boston: Little, Brown, 1929), 3. Mart Stewart made this argument about the ways in which this older tradition of determinist thinking has affected the development of southern environmental history in "'Let Us Begin with the Weather?': Climate, Race, and Cultural Distinctiveness in the American South," in *Nature and Society in Historical Context*, ed. Mikulas Teich, Roy Porter, and Bo Gustaffson (New York: Cambridge University Press, 1997), 240–56. See also Paul S. Sutter, "Introduction: No More the Backward Region: Southern Environmental History Comes of Age," in *Environmental History and the American South: A Reader*, ed. Paul S. Sutter and Christopher Manganiello (Athens: University of Georgia Press, 2009), 1–24.

CHAPTER ONE. Yawning, Abysmal Gullies

1. Jimmy Carter, *Turning Point: A Candidate, a State, and a Nation Come of Age* (New York: Times Books, 1992).

2. U.S. Census Bureau, State and County QuickFacts, Stewart County, Georgia, accessed July 31, 2013, http://quickfacts.census.gov/qfd/states/13/13259.html. On forest cover, see "Community Assessment: Stewart County and the Cities of Lumpkin and Richland," accessed July 31, 2013, http://www.dca.state.ga.us/development/planning qualitygrowth/programs/downloads/plans/StewartAssessmentMain.pdf.

3. William Z. Clark Jr. and Arnold C. Zisa, *Physiographic Map of Georgia* (Atlanta: [Georgia] Department of Natural Resources, 1967), reprinted in "The Fire Forest: Longleaf Pine-Wiregrass Ecosystem," *Georgia Wildlife: Natural Georgia Series* 8, no. 2 (2001): 8–9.

4. Susan Eva O'Donovan, *Becoming Free in the Cotton South* (Cambridge, Mass.: Harvard University Press, 2007), 3. See also Lee W. Formwalt, "A Garden of Irony and Diversity," in *The New Georgia Guide* (Athens: University of Georgia Press, 1996), 500–504.

5. W. E. B. Du Bois, *The Souls of Black Folk: Essays and Sketches* (Chicago: A. C. McClurg, 1904): 113, 122–25.

6. Charles Aiken, *The Cotton Plantation South since the Civil War* (Baltimore: Johns Hopkins University Press, 1998), 10.

7. On these mounds, see M. J. Wood, "Singer-Moye Mounds," New Georgia Encyclopedia, August 29, 2013, http://www.georgiaencyclopedia.org/articles/history-archaeology/singer-moye-mounds, accessed December 18, 2013.

8. Robbie Ethridge, *Creek Country: The Creek Indians and their World* (Chapel Hill: University of North Carolina Press, 2003): 23–31.

9. Ibid., 32–50; Daniel D. Richter, Jr. and Daniel Markewitz, *Understanding Soil Change: Soil Sustainability over Millennia, Centuries, and Decades* (New York: Cambridge University Press, 2001): 107–15.

10. Ethridge, *Creek Country*, 174, 215–19.

11. Sigrid Sanders, "Providence Canyon," New Georgia Encyclopedia, December 13, 2013, http://www.georgiaencyclopedia.org/articles/geography-environment/providence-canyon, accessed October 10, 2014.

12. Robert W. McVety, "Steep-Sided Gully Erosion in Stewart County, Georgia: Causes and Consequences" (master's thesis, Florida State University, 1971), 6–7.

13. See Helen Elisa Terrill, *History of Stewart County, Georgia*, vol. 1, ed. Sara Robertson Dixon (Columbus, Ga.: Columbus Office Supply, 1958), 1–6; Matthew M. Moye, "Stewart County," *New Georgia Encyclopedia*, September 3, 2013, http://www.georgia encyclopedia.org/articles/counties-cities-neighborhoods/stewart-county, accessed December 18, 2013.

14. For 1840 and 1860 data, see Francis J. Magilligan and Melissa L. Stamp, "Historical Land-Cover Changes and Hydrographic Adjustment in a Small Georgia Watershed," *Annals of the American Association of Geographers* 7, no. 4 (1997): 616–18. For 1845 population figures, see George White, *Statistics of the State of Georgia* (Savannah: W. Thorne Williams, 1849), 521. Cotton statistics from Moye, "Stewart County." For 1850 slave numbers, see Terrill, *History of Stewart County, Georgia*, 1:31.

15. Moye, "Stewart County"; Terrill, *History of Stewart County, Georgia*, 1:7.

16. Arthur R. Hall, "Soil Erosion and Agriculture in the Southern Piedmont: A History" (PhD diss., Duke University, 1948), 50–51; David C. Long, M. W. Beck, E. C. Hall, and W. W. Burdette, "Soil Survey of Stewart County, Georgia," in *Field Operations of the Bureau of Soils, 1913* (Washington, D.C.: GPO, 1914), 547.

17. Terrill, *History of Stewart County, Georgia*, 1:174–75.

18. Ibid.

19. Ted Steinberg, *Acts of God: The Unnatural History of Natural Disasters* (New York: Oxford University Press, 2000), xxi.

20. George Perkins Marsh, *Man and Nature; or, Physical Geography as Modified by Human Action* (1864; Seattle: University of Washington Press, 2003). On Marsh, see David Lowenthal, *George Perkins Marsh: Prophet of Conservation* (Seattle: University of Washington Press, 2000).

21. This is an oft-repeated quote. For an example of its deployment, see David R. Montgomery, *Dirt: The Erosion of Civilizations* (Berkeley: University of California Press, 2007), 115.

22. Steven Stoll, *Larding the Lean Earth: Soil and Society in Nineteenth-Century America* (New York: Hill & Wang, 2002).

23. James Bonner and Willard Range both cite a number of Georgia farmers who had achieved model farm status in the antebellum period, David Dickson being a preeminent example. See James C. Bonner, *A History of Georgia Agriculture, 1732–1860* (Athens: University of Georgia Press, 1964), chap. 7; Willard Range, *A Century of Georgia Agriculture, 1850–1950* (Athens: University of Georgia Press, 1954), chap. 1.

24. Albert Lowther Demaree, *The American Agricultural Press, 1819–1860* (New York: Columbia University Press, 1941), 397.

25. See Alan Taylor, "'Wasty Ways': Stories of American Settlement," *Environmental History* 3, no. 3 (July 1998): 291–310.

26. "Stewart County, 1893. R. T. Humber Arranged for a Photographer to Take Photographs of Providence Caves Located near Lumpkin, Ga.," Vanishing Georgia, Georgia Archives, University System of Georgia, http://cdm.georgiaarchives.org:2011/cdm/ref/collection/vg2/id/14490, accessed October 11, 2014.

27. *Lumpkin Independent*, July 16, 1887.

28. "Thanksgiving Day," *Lumpkin Independent*, November 30, 1889. Matthew Moye has done extensive research in Stewart County's local papers and was an invaluable informant for me, bringing stories like these to my attention.

CHAPTER TWO. The Most Picturesque Features of the Coastal Plain

1. Benjamin R. Cohen, *Notes from the Ground: Science, Soil, and Society in the American Countryside* (New Haven: Yale University Press, 2009), 169; Michele L. Aldrich, "American State Geological Surveys, 1820–1845," in Cecil J. Schneer, ed., *Two Hundred Years of Geology in America* (Hanover, N.H.: University Press of New England, 1979): 133–43.

2. James C. Bonner, *A History of Georgia Agriculture, 1732–1860* (Athens: University of Georgia Press, 1964), 111; Leonard G. Wilson, *Lyell in America: Transatlantic Geology, 1841–1853* (Baltimore: Johns Hopkins University Press, 1998), 196.

3. Cohen, *Notes from the Ground*, 166–96. See also Bonner, *History of Georgia Agriculture*, 110.

4. John Ruggles Cotting, *An Essay on the Soils and Available Manures of the State of Georgia, with the Mode of Application and Management Founded on a Geological and Agricultural Survey* (Milledgeville, Ga.: Park & Rogers, 1843), iv–v. Italics in the original.

5. Edmund Ruffin, *Report on the Commencement and Progress of the Agricultural Survey of South-Carolina, for 1843* (Columbia: A. H. Pemberton, 1843); A. R. Hall, *The Story of Soil Conservation in the South Carolina Piedmont, 1800–1860* (Washington, D.C.: U.S. Department of Agriculture, 1940), 5. See also William H. Mathew, ed., *Agriculture, Geology, and Society in Antebellum South Carolina: The Private Diary of Edmund Ruffin, 1843* (Athens: University of Georgia Press, 1992). Ruffin's survey focused on finding deposits of marl to correct for soil acidity.

6. Cohen, *Notes on the Ground*, 167.

7. E. W. Hilgard, *Report on the Geological and Agricultural Survey of the State of Mississippi* (Jackson: Mississippian Steam Power Press, 1858), 239. On Hilgard, see Hans Jenny, *E. W. Hilgard and the Birth of Modern Soil Science* (Pisa, Italy: Collana Della Revista "Agrochimica," 1961); David R. Montgomery, *Dirt: The Erosion of Civilizations* (Berkeley: University of California Press, 2007), 188–92.

8. Steven Stoll, *Larding the Lean Earth: Soil and Society in Nineteenth-Century America* (New York: Hill & Wang, 2002), 162–65.

9. Willard Range, *A Century of Georgia Agriculture, 1850–1950* (Athens: University of Georgia Press, 1954), 121–22; Margaret W. Rossiter, *The Emergence of Agricultural Science: Justus Liebig and the Americans, 1840–1880* (New Haven: Yale University Press, 1975), 149–71.

10. Mott Greene, *Geology in the Nineteenth Century: Changing Views of a Changing World* (Ithaca: Cornell University Press, 1982); Wilson, *Lyell in America*; Martin Rudwick, *Earth's Deep History: How It Was Discovered and Why It Matters* (Chicago: University of Chicago Press, 2014).

11. Gould quoted in David M. Wilkinson, "Ecology before Ecology: Biogeography and Ecology in Lyell's '*Principles*,'" *Journal of Biogeography* 29 (2002): 1109.

12. Barbara A. Kennedy, "Charles Lyell and 'Modern Changes of the Earth': The Milledgeville Gully," *Geomorphology* 40 (2001): 91–98; Wilson, *Lyell in* America, 1–6.

13. See Charles Lyell, *Travels in North America in the Years 1841–42*, 2 vols. (New York: Wiley and Putnam, 1845); Charles Lyell, *A Second Visit to the United States of North America*, 2 vols. (New York: Harper & Brothers, 1849).

14. On Lyell and his travels, see Wilson, *Lyell in America*. Quote is from Kennedy, "Charles Lyell," 92.

15. Lyell, *Second Visit*, 1:256.
16. Wilson, *Lyell in America*, 15–17, 34–35, 270–72.
17. Lyell, *Second Visit*, 2:28–30; Wilson, *Lyell in America*, 196. On efforts to reconstruct the erosional history of Lyell Gully, see H. Andrew Ireland, "'Lyell' Gully, A Record of a Century of Erosion," *Journal of Geology* 47, no. 1 (January–February 1939): 47–63; Kennedy, "Charles Lyell."
18. Lyell, *Second Visit*, 2:35–36; Wilson, *Lyell in America*, 199.
19. Charles Lyell, *Principles of Geology*, 9th ed. (New York: D. Appleton, 1853), 204–5.
20. Barbara Novak, *Nature and Culture: American Landscape and Painting, 1825–1875*, 3rd ed. (1980; New York: Oxford University Press, 2007), 48–53, quote from 50.
21. George Little, *Catalogue of Ores, Rocks, and Woods, Selected from the Geological Survey Collection of the State of Georgia, USA* (Atlanta: Jason P. Harrison, 1878), 14.
22. Donald Worster, *A River Running West: The Life of John Wesley Powell* (New York: Oxford University Press, 2001), 484–85. See also William Goetzmann, *New Lands, New Men: America and the Second Great Age of Discovery* (New York: Viking, 1986).
23. Much of this is taken from Worster, *River Running West*, 484–86.
24. For the history of the Geological Survey of Georgia, see H. S. Cave, *Historical Sketch of the Geological Society of Georgia*, Geological Survey of Georgia Bulletin no. 39 (Atlanta: Foote & Davies, 1922).
25. Otto Veatch, *Second Report on the Clay Deposits of Georgia*, Georgia Geological Survey Bulletin no. 18 (Atlanta: Geological Survey of Georgia, 1909).
26. Otto Veatch and Lloyd William Stephenson, *Preliminary Report on the Geology of the Coastal Plain of Georgia*, Georgia Geological Survey Bulletin no. 26 (Atlanta: Geological Survey of Georgia, 1911), 29–30.
27. Ibid., 30.
28. Stephen J. Pyne, *How the Canyon Became Grand: A Short History* (New York: Penguin, 1998), 20.
29. John Wesley Powell, *The Exploration of the Colorado River of the West and Its Tributaries* (Washington, D.C.: GPO, 1875). This paragraph heavily relies on Pyne, *How the Canyon Became Grand*. On the redemptive West, see William Deverell, "Conquest to Convalescence: Nature and Nation in United States History," in *The Oxford Handbook of Environmental History*, ed. Andrew C. Isenberg (New York: Oxford University Press, 2014), 644–67. For shifting aesthetic tastes during this period, see Novak, *Nature and Culture*; William Cronon, "The Trouble with Wilderness; or, Getting Back to the Wrong Nature," in *Uncommon Ground: Rethinking the Human Place in Nature*, ed. William Cronon (New York: Norton, 1986): 69–90.
30. Pyne, *How the Canyon Became Grand*, 113–16.
31. On the "See America First" movement, see Marguerite Shaffer, *See America First: Tourism and National Identity, 1880–1940* (Washington, D.C.: Smithsonian Institution Press, 2001).

CHAPTER THREE. Rough, Gullied Land

1. For the history of this survey, see Box 222, folder "Georgia, Stewart County, 1913," entry 216—Records of the Soil Survey Division, Reports on Soils Surveys, 1899–1927, record group 54—Records of the Bureau of Plant Industry, Soils, and Agri-

cultural Engineering, National Archives and Records Administration, College Park, Maryland (hereafter NARA II).

2. Douglas Helms, Anne B. Effland, and Patricia J. Durana, eds., *Profiles in the History of the U.S. Soil Survey* (Ames: Iowa State University Press, 2002); David Rice Gardner, *The National Cooperative Soil Survey of the United States* (Washington, D.C.: GPO, 1998); Samuel R. Stalcup, "Public Interest, Private Lands: Soil Conservation in the United States, 1890–1940 (PhD diss., University of Oklahoma, 2014).

3. Helms, Effland, and Durana, *Profiles*, 20–24, 32–34.

4. Humphrey Davy, *Elements of Agricultural Chemistry* (London: Longman, Hurst, Rees, Orme, and Brown, 1813); Steven Stoll, *Larding the Lean Earth: Soil and Society in Nineteenth-Century America* (New York: Hill & Wang, 2002), 151; Gardner, *National Cooperative Soil Survey*, 21–22; Jack Temple Kirby, introduction to *Nature's Management: Writings on Landscape and Reform, 1822–1859*, by Edmund Ruffin, ed. Jack Temple Kirby (Athens: University of Georgia Press, 2000), xxii; Margaret W. Rossiter, *The Emergence of Agricultural Science: Justus Liebig and the Americans, 1840–1880* (New Haven: Yale University Press, 1975), 3–19.

5. Justus von Liebig, *Organic Chemistry and Its Applications to Agriculture and Physiology* (London: Taylor and Walton, 1840). See also Vaclav Smil, *Enriching the Earth: Fritz Haber, Carl Bosch, and the Transformation of World Food Production* (Cambridge, MIT Press, 2001), 8–15; Hugh S. Gorman, *The Story of N: A Social History of the Nitrogen Cycle and the Challenge of Sustainability* (New Brunswick: Rutgers University Press, 2013), 59–63; Rossiter, *Emergence of Agricultural Science*, 10–28; Benjamin R. Cohen, *Notes from the Ground: Science, Soil, and Society in the American Countryside* (New Haven: Yale University Press, 2009), 82–83; Stoll, *Larding the Lean Earth*, 151–53; Kirby, introduction to *Nature's Management*, xxiii–xxv; David R. Montgomery, *Dirt: The Erosion of Civilizations* (Berkeley: University of California Press, 2007), 183–84; Gardner, *National Cooperative Soil Survey*, 22; C. Feller, "The Concept of Soil Humus in the Past Three Centuries," in *History of Soil Science: International Perspectives*, ed. Dan H. Yaalon and S. Berkowicz (Reiskirchen, Germany: Catena, 1997): 15–46.

6. Montgomery, *Dirt*, 184.

7. Rossiter, *Emergence of Agricultural Science*, xii.

8. Douglas Helms, "Early Leaders of the Soil Survey," in Helms, Effland, and Durana, *Profiles*, 22; Rossiter, *Emergence of Agricultural Science*, 127–48; Gardner, *National Cooperative Soil Survey*, 23–25.

9. Smil, *Enriching the Earth*, 11–12.

10. Rossiter, *Emergence of Agricultural Science*, 146.

11. Ibid., 149–71; Ronald Amundson, "Philosophical Developments in Pedology in the United States: Eugene Hilgard and Milton Whitney," in *Footprints in the Soil: People and Ideas in Soil History*, ed. Benno P. Warkentin (Amsterdam: Elsevier, 2006), 149.

12. Helms, "Early Leaders," 20–24; Gardner, *National Cooperative Soil Survey*, 26–29; Montgomery, *Dirt*, 191–92.

13. In 1903, Whitney and F. K. Cameron had laid out the details of Whitney's soil texture theory in *The Chemistry of Soils as Related to Soil Production*, Bureau of Soils Bulletin no. 22 (Washington, D.C.: GPO, 1903).

14. On the Hilgard-Whitney feud, see Amundson, "Philosophical Developments,"

149–65; Helms, "Early Leaders"; Hans Jenny, *E. W. Hilgard and the Birth of Modern Soil Science* (Pisa, Italy: Collana Della Revista "Agrochimica," 1961).

15. More specifically, King showed that nutrients available in soil solutions, and thus available to plants, were quite different in quantity, and better correlated with soil productivity, than nutrient levels revealed in total chemical analysis of soil samples. See Montgomery, *Dirt*, 192; C. B Tanner and R. W. Simonson, "Franklin King—Pioneer Scientist," *Soil Science Society of America Journal* 57, no. 1 (1993): 289–90.

16. John Paull, "The Making of an Agricultural Classic: *Farmers of Forty Acres; or, Permanent Agriculture in China, Korea and Japan*, 1911–2011," *Agricultural Sciences* 2, no. 3 (2011): 175–80; Tanner and Simonson, "Franklin King—Pioneer Scientist," 286–92; F. H. King, *Farmers of Forty Centuries; or, Permanent Agriculture in China, Korea and Japan* (Madison, Wis.: Mrs. F. H. King, 1911); Whitney and Cameron, *Chemistry of Soils*.

17. E. W. Hilgard, "Soil Management," *Science* 20, no. 514 (November 4, 1904): 605–8, quotes from 608.

18. See, for instance, "Report on Statements of Dr. Cyril G. Hopkins Relative to Bureau of Soil," USDA Office of the Secretary Circular No. 22, March 16, 1907.

19. Hopkins quoted in Paull, "Making of an Agricultural Classic," 176.

20. King, *Farmers of Forty Centuries*.

21. Ibid., 48.

22. On this issue, see Alan MacFarlane, *The Savage Wars of Peace: England, Japan and the Malthusian Trap* (Malden, Mass.: Blackwell, 1997), 138–39, 155–58.

23. King, *Farmers of Forty Centuries*, 1. On the history of sewage treatment, see Daniel Schneider, *Hybrid Nature: Sewage Treatment and the Contradictions of the Industrial Ecosystem* (Cambridge: MIT Press, 2011). For this transition in urban America, see Ted Steinberg, "The Death of the Organic City," in *Down to Earth: Nature's Role in American History* (New York: Oxford University Press, 2002), 157–72.

24. On Lowdermilk, see J. Douglas Helms, "Walter Lowdermilk's Journey: Forester to Land Conservationist," *Environmental Review* 8, no. 2 (Summer 1984): 132–45.

25. Cyril G. Hopkins, *Soil Fertility and Permanent Agriculture* (Boston: Ginn, 1910), 313, 340–41; Cyril G. Hopkins, *The Farm That Won't Wear Out* (Champaign, Ill.: C. Hopkins, 1913), 72.

26. Cyril G. Hopkins, *The Story of the Soil: From the Basis of Absolute Science and Real Life*, 6th rev. ed. (Boston: Richard Badger, 1911). On narratives of sectional reconciliation, see David Blight, *Race and Reunion: The Civil War in American Memory* (Cambridge: Harvard University Press, 2001).

27. Paull, "Making of an Agricultural Classic," 175–76.

28. Howard prominently cited King in his most famous manifesto, *An Agricultural Testament* (London: Oxford University Press, 1940).

29. Paull, "Making of an Agricultural Classic," 177.

30. Gardner, *National Cooperative Soil Survey*, 28.

31. These examples are taken from H. H. Bennett, "The Classification of Forest and Farm Lands in the Southern United States," in *Proceedings of the Third Southern Forestry Congress* (New Orleans: John J. Weihing, 1921), 69–70.

32. Gardner, *National Cooperative Soil Survey*, 36–50; Roy W. Simonson, "Evolution of Soil Series and Type Concepts in the United States," in *History of Soil Science:*

International Perspectives, ed. Dan H. Yaalon and S. Berkowicz (Reiskirchen, Germany: Catena Verlag, 1997), 81.

33. Simonson, "Evolution of Soil Series," 83.

34. Gardner, *National Cooperative Soil Survey*, 40–48; Simonson, "Evolution of Soil Series," 92–95; Jay A. Bonsteel and Clarence W. Dorsey, "Soil Survey of Cecil County, Maryland," in *Field Operations of the Division of Soils, 1900* (Washington, D.C.: GPO, 1901), 103–24. On Marbut and the Russian school of soil science, see David Moon, *The Plough that Broke the Steppes: Agriculture and Environment on Russia's Grasslands, 1700–1914* (Oxford, England: Oxford University Press, 2013), 286–87.

35. Gardner, *National Cooperative Soil Survey*, 55–58. Quote on "Yazoo areas" soils is from Milton Whitney, "General Review of the Work," *Field Operations of the Bureau of Soils, 1901* (Washington, D.C.: GPO, 1902), 62; George N. Coffey and Edward Hearn, "Soil Survey of the Cary Area, North Carolina," *Field Operations of the Bureau of Soils, 1901* (Washington, D.C.: GPO, 1902), 314.

36. *Washed Soils: How to Prevent and Reclaim Them*, Farmers' Bulletin no. 20 (Washington, D.C.: GPO, 1894).

37. Hugh H. Bennett, *The Hugh Bennett Lectures* (Raleigh: Agricultural Foundation, North Carolina State College, 1959), 11; Wellington Brink, *Big Hugh: The Father of Soil Conservation* (New York: Macmillan, 1951), 28.

38. Brink, *Big Hugh*, 50–55. On Carver's work in Macon County, see Mark Hersey, *My Work Is That of Conservation: An Environmental Biography of George Washington Carver* (Athens: University of Georgia Press, 2011).

39. Brink, *Big Hugh*, 50–57; Lynn Nelson, *Pharsalia: An Environmental Biography of a Southern Plantation, 1780–1880* (Athens: University of Georgia Press, 2007), 1–4. For Bennett's recollection of the importance of this Louisa County survey, see box 1, folder "Lectures on Soil Erosion: Its Extent and Meaning and Necessary Measures of Control, HHB—11/4/32," entry 21, record group 114, NARA II. For the edited report and correspondence, see box 42, entry 216, two folders titled "Virginia, Louisa County, 1905," record group 54, NARA II. For the final survey, see Hugh H. Bennett and W. E. McLendon, "Soil Survey of Louisa County, Virginia," *Field Operations of the Bureau of Soils, 1905* (Washington, D.C.: GPO, 1906): 191–212.

40. Samuel P. Hays, *Conservation and the Gospel of Efficiency: The Progressive Conservation Movement, 1890–1920* (Cambridge: Harvard University Press, 1959).

41. Donald Worster, *A River Running West: The Life of John Wesley Powell* (New York: Oxford University Press, 2001), 490–91, 536–38.

42. Hays, *Conservation*, 102; Gifford Pinchot, *Breaking New Ground* (1947; Washington, D.C.: Island Press: 1998), 359. Both quoted in Helms, "Early Leaders," 29.

43. Helms, "Early Leaders," 27–29.

44. Bennett, *Hugh Bennett Lectures*, 12–13, Helms, "Early Leaders," 32.

45. Douglas Helms, "The Early Soil Survey: Engine for the Soil Conservation Movement," in Douglas Helms, *He Loved to Carry the Message: The Collected Writings of Douglas Helms* (Raleigh, N.C.: Lulu.com, 2012), 104; W. J. McGee, *Soil Conservation*, Bureau of Soils Bulletin no. 71 (Washington, D.C.: GPO, 1911); Gardner, *National Cooperative Soil Survey*, 70–71.

46. Charles Van Hise, *The Conservation of Natural Resources in the United States* (New York: MacMillan, 1910), 309; *Report of the Country Life Commission* (New York: Sturgis & Walton, 1911), 83–91.

47. R. Burnell Held and Marion Clawson make this point in *Soil Conservation in Perspective* (Baltimore: Johns Hopkins University Press, 1965), 38–40.

48. Milton Whitney, *Soils of the United States, Based upon the Work of the Bureau of Soils to January 1, 1908*, Bureau of Soils Bulletin no. 55 (Washington, D.C.: GPO, 1909), 66–80, quotes from 66, 67. For the Bennett quote, see Brink, *Big Hugh*, 11.

49. *Report of the National Conservation Commission*, 3 vols. (Washington, D.C.: GPO, 1909); see particularly Milton Whitney, "Crop Yield and Soil Composition," 3:9–107, quotes on 35, 36. Ian Tyrrell cites this report and Whitney's presence in it in *Crisis of a Wasteful Nation: Empire and Conservation in Theodore Roosevelt's America* (Chicago: University of Chicago Press, 2015).

50. Letter, Bennett to Dr. Frank Bohn, August 24, 1934, entry 5, box 10, folder 21 (Boa–Bol), p. 4, record group 114, NARA II. I am indebted to Joshua Nygren for sharing this letter with me.

51. Milton Whitney, *Soil and Civilization: A Modern Concept of the Soil and the Historical Development of Agriculture* (New York: D. Van Nostrand, 1925): 89.

52. Bennett, *Hugh Bennett Lectures*, 11–13.

53. Hugh H. Bennett, Howard C. Smith, W. M. Spann, E. M. Jones, and A. L. Goodman, "Soil Survey of Lauderdale County, Mississippi," in *Field Operations of the Bureau of Soils, 1910* (Washington, D.C.: GPO, 1912), 733–84. Bennett et al. quoted in Douglas Helms, "Early Leaders," 32.

54. M. Earl Carr, F. S. Welsh, G. A. Crabb, Risden T. Allen, and W. C. Byers, "Soil Survey of Fairfield County, South Carolina," in *Field Operations of the Bureau of Soils, 1911* (Washington, D.C., GPO: 1912), 505.

55. W. Edward Hearn and Frank Drane, "Soil Survey of Caswell County, North Carolina," in *Field Operations of the Bureau of Soils, 1908* (Washington, D.C., GPO, 1910). For the environmental history of this region, see Drew Swanson, *A Golden Weed: Tobacco and Environment in the Piedmont South* (New Haven: Yale University Press, 2014).

56. Letter, Hugh H. Bennett to Milton Whitney, June 6, 1912 [this may be a typo—I think the letter was from 1911], entry 216, box 157, folder "South Carolina, Fairfield County, 1911," 1, record group 54, NARA II.

57. Report, H. H. Bennett to Milton Whitney, November 28, 1911, entry 216, box 157, folder "South Carolina, Fairfield County, 1911," 1, record group 54, NARA II.

58. Carr et al., "Soil Survey of Fairfield County," 505.

59. Ibid., 510.

60. Letter, David A. Long to chief of the Bureau of Soils, October 26, 1912, entry 216, box 222, folder "Georgia, Stewart County, 1913," record group 54, NARA II.

61. Letter, David A. Long to chief of the Bureau of Soils, May 3, 1913, ibid.

62. Letter, Hugh Hammond Bennett to Milton Whitney, February 13, 1913, ibid.

63. David C. Long, M. W. Beck, E. C. Hall, and W. W. Burdette, "Soil Survey of Stewart County, Georgia," in *Field Operations of the Bureau of Soils, 1913* (Washington, D.C.: GPO, 1914), 561.

64. Ibid., 602, photo opposite p. 554.

65. R. O. E. Davis, "Economic Waste from Soil Erosion," *Yearbook of the Department of Agriculture* (USDA, Washington, D.C.: GPO, 1913), 207–20; R. O. E. Davis, *Soil Erosion in the South*, Bulletin of the U.S. Department of Agriculture no. 180 (Washington, D.C.: GPO, 1915).

66. Davis, *Soil Erosion in the South*, 19.

67. Hugh Hammond Bennett, *The Soils and Agriculture of the Southern States* (New York: Macmillan, 1921), 98.

68. H. H. Bennett, "The Classification of Forest and Farm Lands in the Southern United States," in *Proceedings of the Third Southern Forestry Congress* (New Orleans: John J. Weihing, 1921), 69–113, quotes from 75–76.

69. Brink, *Big Hugh*, 61–71; H. H. Bennett and W. A. Taylor, *The Agricultural Possibilities of the Canal Zone*, USDA Report no. 95 (Washington, D.C.: GPO, 1912); Hugh H. Bennett, *The Soils of Cuba* (Washington, D.C.: Tropical Plant Research Foundation, 1928). On soil conservation during the 1920s, see Donald Swain, *Federal Conservation Policy, 1921–1933* (Berkeley: University of California Press, 1963), 144–50.

70. A. G. McCall, *Report of the Chief of the Bureau of Soils* (Washington, D.C.: GPO, 1927), 2.

PART TWO. Making Providence Canyon Meaningful in the 1930s

1. David L. Carlton and Peter A. Coclanis, eds., *Confronting Poverty in the Great Depression: The Report on Economic Conditions of the South with Related Documents* (Boston: Bedford-St. Martin's, 1991), 42.

2. Sarah Phillips, *This Land, This Nation: Conservation, Rural America, and the New Deal* (New York: Cambridge University Press, 2007), 1.

3. Paul S. Sutter, *Driven Wild: How the Fight against Automobiles Launched the Modern Wilderness Movement* (Seattle: University of Washington Press, 2002), 19–53.

CHAPTER FOUR. A Land That Nature Built for Tourists

1. "Georgia Wonders to Be Described on Constitution News Broadcast," *Atlanta Constitution*, June 25, 1933; "The Grand Canyon of Georgia—No. 11 of the Know Your Georgia Context Series," *Atlanta Constitution*, June 25, 1933, Gravure Pictorial Section.

2. Marguerite Shaffer, *See America First: Tourism and National Identity, 1880–1940* (Washington, D.C.: Smithsonian Institution Press, 2001); Drake Hokanson, *The Lincoln Highway: Main Street across America* (Iowa City: University of Iowa Press, 1999); Richard Starnes, *Creating the Land of the Sky: Tourism and Society in Western North Carolina* (Tuscaloosa: University of Alabama Press, 2010); Margaret Lynn Brown, *The Wild East: A Biography of the Great Smoky Mountains* (Gainesville: University Press of Florida, 2001); Albert G. Way, *Conserving Southern Longleaf: Herbert Stoddard and the Rise of Ecological Land Management* (Athens: University of Georgia Press, 2011); Scott Giltner, *Hunting and Fishing in the New South: Black Labor and White Leisure after the Civil War* (Baltimore: Johns Hopkins University Press, 2008); Mart A. Stewart, *"What Nature Suffers to Groe": Life, Labor, and Landscape on the Georgia Coast, 1680–1920* (Athens: University of Georgia Press, 1996); Tracy J. Revels, *Sunshine Paradise: A History of Florida Tourism* (Gainesville: University Press of Florida, 2011); Nina Silber, *The Romance of Reunion: Northerners and the South, 1865–1900* (Chapel Hill: University of North Carolina Press, 1993).

3. Howard Preston, *Dirt Roads to Dixie: Accessibility and Modernization in the South, 1885–1935* (Knoxville: University of Tennessee Press, 1991). See also Tammy Ingram, *Dixie Highway: Road Building and the Making of the Modern South, 1900–1930* (Chapel Hill: University of North Carolina Press, 2014).

4. "Stewart, Webster, and the FDR Motorcade," *Stewart-Webster Journal* (hereafter *SWJ*), July 12, 1935.

5. "'See Georgia First' Move Endorsed by State Press," *SWJ* (July 5, 1935); "'See Georgia First' Urged by Boosters as Data Is Gathered," *SWJ*, August 16, 1935.

6. "Club Passes Resolutions Urging Park," *Columbus Enquirer* (hereafter *CE*), November 2, 1937.

7. "Stewart to Get CCC Camp for Soil Erosion Project; Will Be Located South of Lumpkin Near Fuller Earth Plant," *SWJ*, May 17, 1935.

8. "The Grand Canyons of Stewart County, Where the Federal Government Will Battle Nature to Preserve Homes and Farm Lands" and "U.S. Soil Erosion Service Fights to Save Georgia Farms from Ruin," *Atlanta Constitution*, September 14, 1935, Gravure Pictorial Section.

9. "Stewart's Famous Caves Illustrated in Government Reports to Show Results of Neglected Lands," *SWJ*, September 20, 1935.

10. See, for instance, "Col. Wimberley Writes Condemning County Erosion Fallacies," *SWJ*, March 12, 1937; "Another Stewart-Born Citizen Resents Big Bug-a-Boo about Caves" *SWJ*, March 12, 1937.

11. "Stewart County's Famous Caves Scenes of Natural Beauty Rather than Horrors of Destructive Element," *SWJ*, May 6, 1937.

12. Editorial, "Make Providence Caves a State Park," *CE*, October 27, 1937.

13. Editorial, "Public-Spirited Lumpkin Working for Park," *CE*, November 5, 1937.

14. "Lumpkin Lions Invite President Roosevelt to Canyons," *CE*, November 2, 1937.

15. "Make Providence Caves a National Park," *CE*, October 27, 1937; "Valley Champion Backs Canyon," *CE*, November 2, 1937.

16. "Lumpkin Lions Club Passes Resolutions Urging Park at 'Little Grand Canyon,'" *SWJ*, November 4, 1937.

17. "Strong Support for Canyon Park," *SWJ*, November 18, 1937.

18. "Club Passes Resolutions Urging Park," *CE*, November 2, 1937.

19. It is not clear to me when this historical marker was erected. The *Stewart-Webster Journal* reported that such a marker was erected in 1940, though the text from which I have quoted comes from a marker that dates to 1953, a year after the Georgia Historical Commission took over the task of erecting historical markers in the state. See Nelson M. Shipp, "Canyons Get a New Marker," *SWJ*, September 12, 1940.

20. Helen Elisa Terrill, *History of Stewart County, Georgia*, ed. Sara Robertson Dixon (Columbus, Ga.: Columbus Office Supply, 1958), 1:385.

21. "Valley Champion Backs Canyon," *CE*, November 2, 1937.

22. Nelson Shipp, "Providence Caves Park Favored by Stewart Leaders," *CE*, October 28, 1937.

23. "Valley Champion Backs Canyon," *CE*, November 2, 1937; "Chamber of Commerce Urges Legislators to Back Park at Canyons," *CE*, November 6, 1937.

24. S. M. Ard, "Canyons Had Largest Crowd Sunday Ever to Visit Colorful Spot," *CE*, November 11, 1937.

25. Providence Canyon Visitor Register, November 17, 1940, Collections of Providence Canyon State Park Visitor Center. In "Head of Georgia Archives Visits Providence Canyons" (*SWJ*, December 15, 1938), the author notes that the visitors were "very much interested in the souvenirs and had painted postcards and took a number home with them."

26. These two terms are used throughout the coverage of the issue by the *Columbus Enquirer* and are also mentioned in Terrill, *History of Stewart County, Georgia*, 1:385.

27. "Visitors Pleased with Canyons," *CE*, November 16, 1937. Shipp made this reference to Bryce Canyon next to his name in the Providence Canyon Visitor Register, July 17, 1938, Collections of Providence Canyon State Park Visitor Center. See also "Government Officials Are Much Impressed after Visit to 'Little Grand Canyons' Tuesday Evening," *SWJ*, February 17, 1938.

28. Nelson M. Shipp, "Fifteen Hundred People See Providence Canyons Sunday," *CE*, November 15, 1937; "Providence Canyons One of the Most Interesting and Impressive Tourist Attractions in Eastern States," *SWJ*, July 7, 1938; "Lions of Zone Hold Meeting in Brink of Providence Canyons," *SWJ*, March 11, 1939.

29. "Visitors Pleased with Canyons," *CE*, November 16, 1937.

30. See "Congressman Plans to Visit Caves," *CE*, November 1, 1937.

31. "Pace Visits Canyon Area," *CE*, November 10, 1937.

32. "Congress Members Support Park," *CE*, November 20, 1937.

33. Robert Fechner to Senator Richard Russell, and Senator Walter George to "Mr. Pike," Columbus Chamber of Commerce, quoted in "National CCC Head Gives Statement upon Canyons Park," *CE*, November 23, 1937.

34. Nelson M. Shipp, "D. A. R. Chapter Urges Little Grand Canyon Be Made a National Park," *CE*, December 13, 1937.

35. See William W. Winn, "The View from Dowdell's Knob," in *The New Georgia Guide* (Athens: University of Georgia Press, 1996): 359–89.

36. "Public-Spirited Lumpkin Working for Park," *CE*, November 5, 1937. Italics are mine. On the Lions Club resolution, see "Lumpkin Lions Club Passes Resolutions Urging Park at 'Little Grand Canyon,'" *SWJ*, November 4, 1937.

37. "Make Providence Caves a National Park," *CE*, October 27, 1937.

38. "Assistant Director of National Park Service Urged to Visit Canyon," *SWJ*, February 10, 1938.

39. "Government Officials Are Much Impressed after Visit to 'Little Grand Canyons' Tuesday Evening," *SWJ*, February 17, 1938.

40. "Canyons Most Colorful Spot in Eastern States, Says Government Report," *SWJ*, April 21, 1938; "Canyons Highly Praised," *SWJ*, April 21, 1938.

41. "Lumpkin Lions Club Welcomes State Authorities at Providence Canyons," *SWJ*, March 10, 1938.

42. "Stewart's Great Opportunity," *SWJ*, September 1, 1938.

43. "Stewart CCC Camp Removed," *SWJ*, September 28, 1939.

44. "Canyons Should Be State Park Site," *SWJ*, July 11, 1940.

45. *Georgia: The WPA Guide to Its Towns and Countryside* (Columbia: University of South Carolina Press, 1990). The photo of "Providence Caverns" appears in the first photo insert, between pages 18 and 19. *Georgia* was originally published by the University of Georgia Press in 1940.

46. *Ibid.*, 452–53.

47. "Darkey of Stewart Author of Poem Book 'Canyons at Providence,'" *SWJ*, February 27, 1941.

48. "Flanagan, Thomas Jefferson," *Who's Who in Colored America*, 7th ed. (Yonkers-on-Hudson, N.Y.: C. E. Burckel, 1950), 187. Thomas Jefferson Flanagan, *The Canyons at Providence (The Lay of the Clay Minstrel)* (Atlanta: Morris Brown College Press, 1940).

49. Cynthia Wachtell, "Sir Walter Scott's Legacy and the Romance of the Civil War," in *War No More: The Antiwar Impulse in American Literature, 1861–1914* (Baton Rouge: Louisiana State University Press, 2010), 32–40.

50. Flanagan, *Canyons at Providence*, 3.

51. Ibid., 4.

52. Ibid., *Canyons at Providence*, 6.

53. Ibid., 8.

54. Ibid., 9.

55. Here I echo David L. Chappell, *A Stone of Hope: Prophetic Religion and the Death of Jim Crow* (Chapel Hill: University of North Carolina Press, 2004).

56. Bernice McCullar, "Tour Guides Ignore Amazing Gully," *Christian Science Monitor*, November 12, 1954.

57. Bern Keating, "Georgia's Little Grand Canyon," *Collier's*, July 22, 1955, 74.

58. Jacquelyn Cook, "Providence Canyon—Exciting Hole in the Ground," *Georgia Magazine* 15, November 5, 1971, 26–27.

59. Information on the creation of the state park has been gleaned from a file of news clippings kept at Providence Canyon State Park. See George Mitchell, "Lumpkin 'Gully' Attracting Lot of Attention," *Columbus Ledger*, March 3, 1969; Beryl Sellars, "Plans Laid for Providence Canyon Park," *Columbus Ledger*, September 5, 1969; "State Plans Major Purchase for Providence Canyon Park," *Columbus Ledger-Enquirer*, November 13, 1970; Beryl Sellers, "Canyon to Become 'Showhouse of Beauty,'" *Columbus Ledger-Enquirer*, January 3, 1971; Luthene Bush, "Canyon Began as a Path," *Albany [Ga.] Herald*, August 5, 1992. See also Robert W. McVety, "Steep-Sided Gully Erosion in Stewart County, Georgia: Causes and Consequences" (master's thesis, Department of Geography, Florida State University, 1971), 110. On Hugh Carter's life, see *Cousin Beedie and Cousin Hot: My Life with the Carter Family of Plains, Georgia* (Englewood Cliffs, N.J.: Prentice-Hall, 1978).

CHAPTER FIVE. Giving Fame and Focus to the Fact of Soil Erosion

1. These photographs can be found by going to the "America from the Great Depression to World War Two: Black-and-White Photographs from the FSA-OWI, 1934–1945" collection, which is part of the American Memory section of the Library of Congress website: http://memory.loc.gov/ammem/fsahtml/fahome.html. A search for "Stewart County" will retrieve the photos.

2. On the documentary impulse, see William Stott, *Documentary Expression and Thirties America* (New York: Oxford University Press, 1973).

3. I borrow the term "New Conservation" from Sarah Phillips, *This Land, This Nation: Conservation, Rural America, and the New Deal* (New York: Cambridge University Press, 2007).

4. On "soil jeremiads," see Randal S. Beeman and James A. Pritchard, *A Green and Permanent Land: Ecology and Agriculture in the Twentieth Century* (Lawrence: University Press of Kansas, 2001), 11.

5. Letter, Hugh Hammond Bennett to George Horace Lorimer, March 29, 1934, box 1, folder 16, Bennett Papers, Conservation Collection, Denver Public Library (hereafter DPL).

6. On relation between Mississippi floods and soil conservation, see Samuel R.

Stalcup, "Public Interest, Private Lands: Soil Conservation in the United States, 1890–1940 (PhD diss., University of Oklahoma, 2014), 135–43.

7. H. H. Bennett, "Uncle Sam Spendthrift," *Scientific American*, April 1927, 237–39; transcript in box 1, folder 3, Bennett Papers, DPL.

8. Hugh Hammond Bennett, "Lectures before the USDA Graduate School," *Soil Conservation Service Technical Paper* 7 (1928): lecture 1, 6–9; lecture 2, 8–11.

9. H. H. Bennett, "The Geographical Relation of Soil Erosion to Land Productivity," *Geographical Review* 18 (October 1928): 579–605.

10. Hugh Hammond Bennett, "Our Vanishing Farm Lands," *North American Review*, August 1929, 171.

11. Douglas Helms, "Hugh Hammond Bennett and the Creation of the Soil Erosion Service," *Journal of Soil and Water Conservation* 64, no. 2 (2009): 68A.

12. H. H. Bennett and W. R. Chapline, *Soil Erosion, a National Menace* (Washington, D.C.: GPO, 1928), 6–7.

13. Ibid., 7.

14. Wellington Brink, *Big Hugh: The Father of Soil Conservation* (New York: Macmillan, 1951), 77–81; Helms, "Hugh Hammond Bennett," 68A.

15. Bennett, "Soil Erosion a Costly Farm Evil," speech presented at Ohio State University, manuscript in box 1, folder 14, Bennett Papers, DPL.

16. H. H. Bennett, "Battling the Rains," box 1, folder 16, Bennett Papers, DPL. This manuscript was attached to a letter from Bennett to George Horace Lorimer, the editor of the *Saturday Evening Post*, where Bennett hoped to get the piece published. But it does not appear that he met with success.

17. Bennett, "The Menace of Erosion," manuscript in box 1, folder 17, Bennett Papers, DPL.

18. Helms, "Hugh Hammond Bennett."

19. For basic background on the National Resources Committee and its predecessors and successors, see Phillips, *This Land, This Nation*, 112–13; A. L. Reisch-Owen, *Conservation under F.D.R.* (New York: Praeger, 1983), 146–49. For the full report, see *A Report on National Planning and Public Works in Relation to Natural Resources and including Land Use and Water Resources with Findings and Recommendations* (Washington, D.C.: GPO, 1934), quote on p. v; map appears as insert between pages 170 and 171. On the Erosion Reconnaissance Survey, see "National Erosion Reconnaissance," section 4 of *Soil Erosion: A Critical Problem in American Agriculture*, Supplementary Report of the Land Planning Committee to the National Resources Board, part 5 (Washington, D.C.: GPO, 1935), 19–22. The full-color version of the "Reconnaissance Erosion Survey" map appears as an insert at the end of this report.

20. On Navajo herd reductions, see Marsha Weisiger, *Dreaming of Sheep in Navajo Country* (Seattle: University of Washington Press, 2009).

21. The Bennett quote is from Jonathan Daniels, *Tar Heels: A Portrait of North Carolina* (New York: Dodd, Mead, 1941), 188; the other quote is from Brink, *Big Hugh*, 7.

22. Most of the information from this paragraph is drawn from Douglas Helms, "Hugh Hammond Bennett and the Creation of the Soil Conservation Service, September 19, 1933–April 27, 1935," *Historical Insights*, no. 9 (March 2010): 1–25.

23. Donald Worster, *Dust Bowl: The Southern Plains in the 1930s* (New York: Oxford University Press, 1979), 213.

24. Hugh H. Bennett, "Soil Loss through Erosion Threatens 'Our Basic Asset,'" *New York Times*, June 17, 1934.

25. H. H. Bennett, "Soil Erosion—A National Menace," *Scientific American*, November 1934, reprinted in *Soil-Erosion Program: Hearing before a Subcommittee of the Committee on the Public Lands*, House of Representatives, 74th Cong., first sess., March 20–22, 25, 1935 (Washington, D.C.: GPO, 1935), 37–45, quote from 39.

26. Hugh H. Bennett and W. C. Lowdermilk, "General Aspects of the Soil-Erosion Problem," *Soils and Men*, USDA Yearbook of Agriculture (Washington, D.C.: GPO, 1938), 587–88.

27. Hugh Hammond Bennett, *Soil Conservation* (New York: McGraw-Hill, 1939), 4.

28. Hugh Hammond Bennett, *Elements of Soil Conservation*, 2nd ed. (New York: McGraw-Hill, 1955), 10.

29. Bennett, *Soil Conservation*, 56–58.

30. Ibid., 67. Figure 27, another image of "Providence Cave," appears on p. 66. Bennett would also feature Providence Canyon in Hugh Hammond Bennett and William Clayton Pryor, *The Land We Defend* (New York: Longmans, Green, 1942).

31. Randall Beeman, "Friends of the Land and the Rise of Environmentalism, 1940–1954," *Journal of Agricultural and Environmental Ethics* 8, no. 1 (1995): 1–16; Neil Maher, *Nature's New Deal: The Civilian Conservation Corps and the Roots of the American Environmental Movement* (New York: Oxford University Press, 2008), 209–10.

32. For Lord's biography, see Finis Dunaway, *Natural Visions: The Power of Images in American Environmental Reform* (Chicago: University of Chicago Press, 2005), 97–99. See also Russell and Kate Lord, *Forever the Land: A Country Chronicle and Anthology* (New York: Harper, 1950).

33. George Mitchell, "Lumpkin 'Gully' Attracting Lot of Attention," *Columbus Ledger*, March 3, 1969. See also Lloyd Williams, "Providence Canyon: Georgia's Big Beautiful Gully," *Georgia Life*, Autumn 1975, 23–24.

34. Luthene Bush, "Canyon Began as a Path," *Albany (Ga.) Herald*, August 5, 1992; Tom Ham, "Wasteful, Beautiful Providence Canyons: Where Cows Going to Water Made Georgia a Wonderland," *Atlanta Journal*, November 25, 1941.

35. Russell Lord, *Behold This Land* (Boston: Houghton-Mifflin, 1938), photograph in insert after 48.

36. Russell Lord, *To Hold This Soil*, Miscellaneous Publication no. 321 (Washington, D.C.: USDA, 1938): 49–50.

37. Quincy C. Ayres, *Soil Erosion and Its Control* (New York: McGraw-Hill, 1936), 6.

38. Austin Earle Burges, *Soil Erosion Control: A Practical Exposition of the New Science of Soil Conservation for Students, Farmers, and the General Public* (Atlanta: Turner E. Smith, 1937), 7–8.

39. William van Dersal, *This American Land: Its History and Uses* (New York: Oxford University Press, 1943), plate 59 in insert opposite 192.

40. Karl B. Mickey, *Man and Soil: A Brief Introduction to the Study of Soil Conservation* (Chicago: International Harvester, 1945), 33.

41. Vance T. Holliday, "A History of Soil Geomorphology in the United States," in *Footprints in the Soil: People and Ideas in Soil History*, ed. Benno Warkentin (Amsterdam: Elsevier, 2006), 193.

42. F. Grave Morris, "Soil Erosion in South-Eastern United States," *The Geographical Journal* 90, no. 4 (October 1937): 363–70, quotes from 368.

43. G. V. Jacks and R. O. Whyte, *Vanishing Lands: A World Survey of Soil Erosion* (New York: Doubleday, Duran, 1939), 14–15. *Vanishing Lands* was originally published as a technical bulletin titled "Erosion and Soil Conservation," and it appeared in Britain under the title *The Rape of the Earth*. See also Kate B. Showers, "Soil Erosion and Conservation: An International History and a Cautionary Tale," in Warkentin, ed., *Footprints in the Soil*, 386.

44. Leon F. Sisk, "All This Started from the Trickle from a Roof," *Soil Conservation* 1, no. 2 (September 1935): 12–13.

45. Wellington Brink, "Some Cracking Big Gullies," *Soil Conservation* 2, no. 12 (June 1937): 282–84, quotes on 283. For other examples of scs use of Providence Canyon, see *Saving Georgia's Soils* (Atlanta: M. D. Collins, 1938), 20; *The Tobacco-and-Cotton South*, USDA SCS Miscellaneous Publication no. 474 (Washington, D.C.: GPO, 1941), 5.

46. Glenda Elizabeth Gilmore, *Gender and Jim Crow: Women and the Politics of White Supremacy in North Carolina, 1860–1920* (Chapel Hill: University of North Carolina Press, 1996): 64–66, 83–84, 88–89.

47. On Daniels, see Charles Eagles, *Jonathan Daniels and Race Relations: The Evolution of a Southern Liberal* (Knoxville: University of Tennessee Press, 1982); Patricia Sullivan, *Days of Hope: Race and Democracy in the New Deal Era* (Chapel Hill: University of North Carolina Press, 1996), 66–67.

48. Jonathan Daniels, *A Southerner Discovers the South* (New York: MacMillan, 1938), 299.

49. Ibid., 302.

50. Ibid., 303–5.

51. Ibid., 302.

52. Ibid., 305.

53. Harold P. Henderson, "Ellis Arnall (1907–1992)," New Georgia Encyclopedia, September 9, 2014, http://www.georgiaencyclopedia.org/articles/government-politics/ellis-arnall-1907-1992, accessed October 14, 2014.

54. Ellis Gibbs Arnall, *The Shore Dimly Seen* (Philadelphia: J. B. Lippincott, 1946), 63.

55. Ibid., 80.

56. Henderson, "Ellis Arnall."

57. See Robert Westbrook, "Tribune of the Technostructure: The Popular Economics of Stuart Chase," *American Quarterly* 32, no. 4 (Fall 1980): 387–408; William Alan Hodson and John Carfora, "Stuart Chase: Brief Life of a Public Thinker, 1888–1985," *Harvard Magazine* (September–October 2004), http://www.harvardmagazine.com/on-line/090431.html, accessed October 14, 2014; Arthur Schlesinger, *The Crisis of the Old Order, 1919–1933* (Boston: Houghton Mifflin, 1957), 403.

58. Stuart Chase, *Rich Land, Poor Land: A Study of Waste in the Natural Resources of America* (New York: Whittlesey House, 1936).

59. Stuart Chase, "Where the Crop Lands Go," *Harper's* 173 (August 1936): 225–33.

60. Chase, *Rich Land, Poor Land*, 92–93.

61. Ibid., 94.

62. Ibid., 94–95.

63. Ibid., 94–96.

64. Ibid., 98–99.

65. Stott, *Documentary Expression and Thirties America*, 9.

66. There is a large literature on Flaherty and the history behind the making of *The Land*. Finis Dunaway discusses the film at length in *Natural Visions*. In particular, see his chapter on *The Land*, titled "A Flicker of Permanence," 87–113. See also Richard Griffith, *The World of Robert Flaherty* (New York: Duell, Brown, and Prince, 1953); Arthur Calder-Marshall, *The Innocent Eye: The Life of Robert J. Flaherty* (New York: Harcourt, Brace, & World, 1963), 185–200; Mike Weaver, *Robert Flaherty's The Land*, American Arts Pamphlet No. 5 (Exeter, UK: University of Exeter, 1979); Paul Rotha, *Robert J. Flaherty: A Biography* (Philadelphia: University of Pennsylvania Press, 1983); and Richard Barsam, *The Vision of Robert Flaherty: The Artist as Myth and Filmmaker* (Bloomington: Indiana University Press, 1988).

67. Dunaway, *Natural Visions*, 88–89; Griffith, *World of Robert Flaherty*, 139.

68. Barsam, *Vision of Robert Flaherty*, 94.

69. Dunaway, *Natural Visions*, 99–100.

70. Russell Lord, "Out of Eden," in *Forever the Land: A Country Chronicle and Anthology*, ed. Russell Lord and Kate Lord (New York: Harper & Brothers, 1950), 25–26.

71. Russell Lord, "With Flaherty Afield," in Lord and Lord, *Forever the Land*, 31–32.

72. Griffith, *World of Robert Flaherty*, 140–41; Calder-Marshall, *Innocent Eye*, 189–99; Rotha, *Robert J. Flaherty*, 221–22; Dunaway, *Natural Visions*, 99–110.

73. Grierson quoted by Calder-Marshall, *Innocent Eye*, 199. Se also Dunaway, 108–10; Rotha, *Robert J. Flaherty*, 221–222; Griffith, *World of Robert Flaherty*, 141.

74. Dunaway, *Natural Visions*, 90.

75. For oblique mentions by environmental historians, see Martin Melosi, "Environment," in *Encyclopedia of Southern Culture*, ed. Charles Reagan Wilson and William Ferris (Chapel Hill: University of North Carolina Press, 1989), 319; Ted Steinberg, *Down to Earth: Nature's Role in American History* (New York: Oxford University Press, 2002), 85; Jack Temple Kirby, *Mockingbird Song: Ecological Landscapes of the South* (Chapel Hill: University of North Carolina Press, 2006), 96. For postwar appearances in textbooks used in various soil science classes, and in studies by geographers, soil scientists, hydrologists, and geomorphologists, see Glenn O. Schwab, Richard K. Frevert, Kenneth K. Barnes, and Talcott W. Edminster, *Elementary Soil and Water Engineering* (New York: John Wiley and Sons, 1957), 86; Kermit C. Berger, *Introductory Soils* (New York: Macmillan, 1965), 350; Kermit C. Berger, *Sun, Soil, and Survival: An Introduction to Soils* (Norman: University of Oklahoma Press, 1972), 350; Miloš Holý, *Erosion and Environment* (New York: Pergamon Press, 1980), 25, 29.

76. Phillips, *This Land, This Nation*, 242–83.

77. On these developments, see Pete Daniel's work, particularly *Lost Revolutions: The South in the 1950s* (Chapel Hill: University of North Carolina Press, 2000); *Toxic Drift: Pesticides and Health in the Post–World War Two South* (Baton Rouge: Louisiana State University Press, 2005); and *Dispossession: Discrimination against African American Farmers in the Age of Civil Rights* (Chapel Hill: University of North Carolina Press, 2013). See also Alan Brinkley, *The End of Reform: New Deal Liberalism in Recession and War* (New York: Knopf, 1995); Phillips, *This Land, This Nation*.

78. See Adam Rome, *The Bulldozer in the Countryside: Suburban Sprawl and the Rise of American Environmentalism* (New York: Cambridge University Press, 2001), 200–209.

79. Daniel, *Dispossession*.

80. Edward Higbee, *The American Oasis: The Land and Its Uses* (New York: Alfred A. Knopf, 1957), 35, 212.

81. Edward Higbee, *The Squeeze: Cities without Space* (New York: Morrow, 1960). On Higbee and his role as an urban environmental critic, see Rome, *The Bulldozer in the Countryside*, 97–98, 142–43, 176–77.

82. Rebecca Gibson, "Canyon Creation," *Creation ex Nihilo* (September–November 2000): 46–48. See also Emmett L. Williams, "Providence Canyon, Stewart County, Georgia—Evidence of Recent Rapid Erosion," *Creation Research Science Quarterly* 32 (June 1995): 29–43.

CHAPTER SIX. Gullies and What They Mean

1. The classic treatment of the environmental dimensions of the Dust Bowl is Donald Worster, *Dust Bowl: The Southern Plains in the 1930s* (New York: Oxford University Press, 1979). Geoff Cunfer has challenged Worster's narrative, arguing that the Dust Bowl was a natural disaster, that the worst erosion was on lands that had never been plowed, and that agriculture on the Great Plains has been surprisingly stable since 1920. See Geoff Cunfer, *On the Great Plains: Agriculture and Environment* (College Station: Texas A&M University Press, 2005).

2. The survey was not perfect; it was completed quickly and at a broad resolution, and some have argued that it exaggerated the problem. See R. Burnell Held and Marion Clawson, *Soil Conservation in Perspective* (Baltimore: Johns Hopkins University Press, 1965), 61.

3. *Soil Erosion: A Critical Problem in American Agriculture*, Supplementary Report of the Land Planning Committee to the National Resources Board, part 5 (Washington, D.C.: GPO, 1935), 23–24.

4. H. H. Bennett, *Soil-Erosion Program: Hearing before a Subcommittee of the Committee on the Public Lands*, House of Representatives, 74th Cong., first sess., March 20–22, 25, 1935, (Washington, D.C.: GPO, 1935), 3–4.

5. Howard W. Odum, *Southern Regions of the United States* (Chapel Hill: University of North Carolina Press, 1936), 38–39.

6. Stanley W. Trimble, *Man-Induced Soil Erosion on the Southern Piedmont, 1700–1970*, enhanced ed. (Ankeny, Iowa: Soil and Water Conservation Society, 2008), 1; H. H. Bennett, "Cultural Changes in Soils from the Standpoint of Erosion," *Journal of the American Society of Agronomy* 23 (1931): 436.

7. Bennett, "Cultural Changes," 437–38.

8. Ibid., 438–40. On Lowndes County, see Hassan Jeffries, *Bloody Lowndes: Civil Rights and Black Power in Alabama's Black Belt* (New York: NYU Press, 2009).

9. C. R. Jackson, J. K. Martin, D. S. Leigh, and L. T. West, "A Southeastern Piedmont Watershed Sediment Budget," *Journal of Soil and Water Conservation* 60, no. 6 (2005): 298–310.

10. Edmund Ruffin, "On Reclamation—Against Erosion: On Draining: Addressed to Young Farmers" (1833–34), in *Nature's Management: Writings on Landscape and Reform, 1822–1859*, by Edmund Ruffin, ed. Jack Temple Kirby (Athens: University of Georgia Press, 2000), 176.

11. Bennett, "Cultural Changes," 436.

12. Stanley Trimble, "Culturally Accelerated Sedimentation on the Middle Georgia Piedmont" (master's thesis, University of Georgia, 1969), 29–33.

13. Stafford C. Happ, "Sedimentation in South Carolina Piedmont Valleys," *American Journal of Science* 243, no. 3 (March 1945): 126. On Happ's importance, see Trimble, "Culturally Accelerated Sedimentation," 6.

14. Stafford C. Happ, Gordon Rittenhouse, and G. C. Dobson, *Some Principles of Accelerated Stream and Valley Sedimentation*, USDA Technical Bulletin no. 695 (Washington, D.C.: GPO, 1940).

15. Ibid., 89.

16. Arthur R. Hall, *The Story of Soil Conservation in the South Carolina Piedmont, 1800–1860*, USDA Miscellaneous Publication no. 407 (Washington, D.C.: GPO, 1940), 11. On the history of malaria in the South, see Margaret Humphreys, *Malaria: Poverty, Race, and Public Health in the United States* (Baltimore: Johns Hopkins University, 2001). Humphreys does not discuss this connection between erosion and malaria in the South, but John R. McNeill does discuss it in the Caribbean context in *Mosquito Empires: Ecology and War in the Greater Caribbean, 1620–1914* (New York: Cambridge University Press, 2010), 28.

17. Jackson et al., "A Southeastern Piedmont Watershed Sediment Budget," 309.

18. On the history of Scull Shoals, see E. Merton Coulter, "Scull Shoals: An Extinct Georgia Manufacturing and Farming Community," *Georgia Historical Quarterly* 48, no. 1 (March 1964): 33–63; Bruce K. Ferguson, "The Alluvial History and Environmental Legacy of the Abandoned Scull Shoals Mill," *Landscape Journal* 18, no. 2 (1999): 147–56; Bruce K. Ferguson, "The Alluvial Progress of Piedmont Streams," in *Effects of Watershed Development and Management on Aquatic Ecosystems*, ed. Larry A. Roesner (New York: American Society of Civil Engineers, 1997), 133–34; Arthur Raper, *Tenants of the Almighty* (New York: Macmillan, 1943); Jonathan Bryant, *How Curious a Land: Conflict and Change in Greene County, Georgia, 1850–1885* (Chapel Hill: University of North Carolina Press, 1996); H. Harold Brown, *The Greening of Georgia: The Improvement of the Environment in the Twentieth Century* (Macon: Mercer University Press, 2002), 71–73, 92–96. On Nathanael Greene, see Natalie D. Saba, "Nathanael Greene (1742–1786)," New Georgia Encyclopedia, September 16, 2014, http://www.georgiaencyclopedia.org/articles/history-archaeology/nathanael-greene-1742-1786, accessed November 6, 2014.

19. Andrews quoted in Raper, *Tenants of the Almighty*, 54–55.

20. Ibid., 55.

21. Bryant, *How Curious a Land*, 171–72; Coulter, "Scull Shoals," 45–55.

22. Quoted in Ferguson, "Alluvial History," 152.

23. Ibid., 133.

24. Ibid., "Alluvial History," 150–52. Robert C. Walter and Dorothy J. Merritts argue that most Piedmont rivers and streams today are artifacts of a huge number of small dams and the trapping of sediments behind them. While they have focused their studies on Pennsylvania and Maryland, the same processes likely occurred in the deeper South. See Robert C. Walter and Dorothy J. Merritts, "Natural Streams and the Legacy of Water-Powered Mills," *Science* 319 (January 2008): 299–304.

25. Ferguson, "Alluvial History," 154, 155.

26. Trimble, *Man-Induced Soil Erosion*, 75–76; Hugh Hammond Bennett, *Soil Conservation* (New York: McGraw-Hill, 1939), 277.

27. Trimble, *Man-Induced Soil Erosion*, vi.

28. S. W. McCallie, *A Preliminary Report on Drainage Reclamation in Georgia*, Geological Survey of Georgia Bulletin no. 25 (Atlanta: Foote & Davies, 1911).

29. H. H Barrows, J. V. Phillips, and J. E. Brantly, *Agricultural Drainage in Georgia*, Geological Survey of Georgia Bulletin no. 32 (Atlanta: Byrd Printing, 1917), quote from p. 4; drainage district statistics from p. xii. See also Brown, *Greening of Georgia*, 42–43.

30. Ferguson, "Alluvial Progress," 132.

31. Lewis Cecil Gray, *History of Agriculture in the Southern United States to 1860*, 2 vols. (1933; Gloucester, Mass.: Peter Smith, 1958), vol. 1:447.

32. The phrase "vegetative blanket" is borrowed from Arthur R. Hall, "Soil Erosion and Agriculture in the Southern Piedmont: A History" (PhD diss., Duke University, 1948), 16.

33. Silver, *A New Face on the Countryside: Indians, Colonists, and Slaves in South Atlantic Forests, 1500–1800* (New York: Cambridge University Press, 1990), 114–15.

34. Hall, "Soil Erosion," 50–51, 71–72. On the relationship between cane and fertility, see Mart Stewart, "From King Cane to King Cotton: Razing Cane in the Old South," *Environmental History* 12 (January 2007): 59–79.

35. A. R. Hall, "Early Erosion Control Practices in Virginia," USDA, Miscellaneous Publication no. 256 (Washington, D.C.: GPO, 1937), 2; Gray, *History of Agriculture*, 1:448; Hall, *The Story of Soil Conservation*, 3; Ulrich Bonnell Phillips, *Life and Labor in the Old South* (1929; Boston: Little, Brown, 1935), 112

36. See Paul W. Gates, *The Farmer's Age: Agriculture 1815–1860* (New York: Holt, Rinehart, and Winston, 1960), 4; Albert E. Cowdrey, *This Land, This South: An Environmental History*, rev. ed. (Lexington: University Press of Kentucky, 1996), 77.

37. James C. Bonner, *A History of Georgia Agriculture, 1732–1860* (Athens: University of Georgia Press), 97. Willard Range also makes this point in *A Century of Georgia Agriculture, 1850–1950* (Athens: University of Georgia Press, 1954), 19.

38. Hall, "Early Erosion Control Practices," 15–20; Hall, "Soil Erosion," 270–75; Bonner, *History of Georgia Agriculture*, 97; John Hebron Moore, *Agriculture in Antebellum Mississippi* (1958; Columbia: University of South Carolina Press, 2010), 44–45; Avery Odelle Craven, *Soil Exhaustion as a Factor in the Agricultural History of Virginia and Maryland, 1606–1860* (1926; Gloucester, Mass.: Peter Smith, 1965), 90–91.

39. Moore, *Agriculture in Antebellum Mississippi*, 45.

40. Hall, "Early Erosion Control Practices," 20.

41. This paragraph is heavily reliant on the work of James C. Bonner, *History of Georgia Agriculture*, 93–109, quote from 100. See also Moore, *Agriculture in Antebellum Mississippi*, 177; Hall, "Early Erosion Control Practices," 23; Hall, "Soil Erosion," 275–79.

42. Willard Range, *A Century of Georgia Agriculture, 1850–1950* (Athens: University of Georgia Press, 1954), 119; Trimble, *Man-Induced Soil Erosion*, 58; Stanley W. Trimble, "Perspectives on the History of Erosion Control in the Eastern United States," in *The History of Soil and Water Conservation*, ed. Douglas Helms and Susan L. Flader (Washington, D.C.: Agricultural History Society, 1985): 61–62.

43. Wellington Brink, *Big Hugh: The Father of Soil Conservation* (New York: Macmillan, 1951), 28.

44. H. A. Ireland, C. F. S. Sharpe, and D. H. Eargle, *Principles of Gully Erosion in*

the *Piedmont of South Carolina*, USDA Technical Bulletin no. 633 (Washington, D.C.: GPO: 1939), 42.

45. See Trimble, *Man-Induced Soil Erosion*, appendix B, 58–59; Hall, "Early Erosion Control Practices."

46. Chris Evans, "The Plantation Hoe: The Rise and Fall of an Atlantic Commodity, 1650–1850," *William and Mary Quarterly* 69, no. 1 (January 2012): 71–100.

47. Madison quoted in Craven, *Soil Exhaustion*, 36.

48. Ibid., 89–92, 111, 152; Bonner, *History of Georgia Agriculture*, 93–97; Hall, "Soil Erosion," 51.

49. Gray, *History of Agriculture*, 1:446.

50. Craven, *Soil Exhaustion*, 19. On Craven, see Jack Temple Kirby, *Mockingbird Song: Ecological Landscapes of the South* (Chapel Hill: University of North Carolina Press, 2006), 88–91. On the ubiquity of the phrase "worn out lands," see Margaret W. Rossiter, *The Emergence of Agricultural Science: Justus Liebig and the Americans, 1840–1880* (New Haven: Yale University Press, 1975) 3–9.

51. Paul W. Gates, *The Farmer's Age: Agriculture, 1815–1860* (New York: Holt, Rinehart, and Winston, 1960), 9.

52. Bonner, *History of Georgia Agriculture*, 32–46; Jim Gigantino, "Land Lottery System," New Georgia Encyclopedia, June 5, 2014, http://www.georgiaencyclopedia.org/articles/history-archaeology/land-lottery-system, accessed October 21, 2014; Farris W. Cadle, *Georgia Land Surveying History and Law* (Athens: University of Georgia Press, 1991); Frederick B. Gates, "The Georgia Land Act of 1803: Political Struggle in a One-Party State," *Georgia Historical Quarterly* 82 (Spring 1998): 1–21.

53. Sources in this paragraph quoted in Bonner, *History of Georgia Agriculture*, 60–64.

54. Jack Temple Kirby, *Poquosin: A Study of Rural Landscape and Society* (Chapel Hill: University of North Carolina Press, 1995), 103–11; Lois Green Carr, Russell R. Menard, and Lorena Walsh, *Robert Cole's World: Agriculture and Society in Early Maryland* (Chapel Hill: University of North Carolina Press, 1991); Steven Stoll, *Larding the Lean Earth: Soil and Society in Nineteenth-Century America* (New York: Hill & Wang, 2002), 130–31; John Majewski, *Modernizing a Slave Economy: The Economic Vision of the Confederate Nation* (Chapel Hill: University of North Carolina Press, 2009), 22–52; Carville Earle, "The Myth of the Southern Soil Miner: Macrohistory, Agricultural Innovation, and Environmental Change," in *The Ends of the Earth: Perspectives on Modern Environmental History*, ed. Donald Worster (New York: Cambridge University Press, 1988), 175–210.

55. Stoll, *Larding the Lean Earth*, 122–23, 128.

56. Majewski, *Modernizing a Slave Economy*, 22–52.

57. Kirby, *Poquosin*, 105–11.

58. Majewski, *Modernizing a Slave Economy*, 53–80; Stoll, *Larding the Lean Earth*, 120–69.

59. Jefferson quoted in William Chandler Bagley, *Soil Exhaustion and the Civil War* (Washington, D.C.: American Council on Public Affairs, 1942), 70; Craven, *Soil Exhaustion*, 34. See also Stoll, *Larding the Lean Earth*, 49–68.

60. The literature on the southern open range is quite large. For the early history of the open range in the South, see Virginia Anderson, *Creatures of Empire: How Domestic Animals Transformed Early America* (New York: Oxford University Press, 2004):

107–40; Craven, *Soil Exhaustion*, 33–34. For the postbellum period, when the open range closed, see Claire Strom, *Making Catfish Bait out of Government Boys: The Fight against Cattle Ticks and the Transformation of the Yeoman South* (Athens: University of Georgia Press, 2009); Ted Steinberg, *Down to Earth: Nature's Role in American History*, 2nd ed. (New York: Oxford University Press, 2009), 105–10; Steven Hahn, *The Roots of Southern Populism: Yeomen Farmers and the Transformation of the Georgia Upcountry, 1850–1890* (New York: Oxford University Press, 1983); Hahn, "Hunting, Fishing, and Foraging: Common Rights and Class Relations in the Postbellum South," *Radical History Review* 26 (1982): 37–64; J. Crawford King, "The Closing of the Southern Range: An Exploratory Study," *Journal of Southern History* 48 (February 1982): 53–70; Shawn Everett Kantor and J. Morgan Kousser, "Common Sense or Commonwealth? The Fence Law and Institutional Change in the Postbellum South," *Journal of Southern History* 59 (May 1993): 201–42.

61. Anderson, *Creatures of Empire*, 117–23; Hall, "Soil Erosion," 49.

62. Stoll, *Larding the Lean Earth*, 126–28.

63. See Craven, *Soil Exhaustion*, especially 10–24; Hall, "Soil Erosion," 51.

64. Stanley Trimble, "Perspectives." See also Trimble's appendix C, "The Importance of Abandoned Land to Erosive Land Use," in *Man-Induced Soil Erosion*, 153–56.

65. R. O. E. Davis, *Soil Erosion in the South*, Bulletin of the U.S. Department of Agriculture no. 180 (Washington, D.C.: GPO, 1915), 21.

66. Raper, *Tenants of the Almighty*, 158–59.

67. For a good discussion of these sources, see Trimble, *Man-Induced Soil Erosion*, 60–61.

68. Lorena S. Walsh, "Land Use, Settlement Patterns, and the Impact of European Agriculture, 1620–1820," in *Discovering the Chesapeake: The History of an Ecosystem*, ed. Philip D. Curtin, Grace S. Brush, and George W. Fisher (Baltimore: Johns Hopkins University Press, 2001), 238.

69. Earle, "Myth."

70. Ibid., 195.

71. Ibid., 192–200; Walsh, "Land Use"; Craven, *Soil Exhaustion*; Carville Earle and Ronald Hoffman, "Genteel Erosion: The Ecological Consequences of Agrarian Reform in the Chesapeake, 1730–1840," in Curtin, Brush, and Fisher, *Discovering the Chesapeake*, 279–303.

72. Earle, "Myth," 201.

73. Ibid., 201–4. Earle relied on the work of John Hebron Moore and Sam Bowers Hilliard. Eugene Genovese suggested that Earle overestimated the extent of this rotation system. See Moore, *Agriculture in Antebellum Mississippi*, 59–60; Sam Bowers Hilliard, *Hog Meat and Hoecake: Food Supply in the Old South, 1840–1860* (Carbondale: Southern Illinois University Press, 1972), 178; Eugene Genovese, "Cotton, Slavery, and Soil Exhaustion in the Old South," *Cotton History Review* 2, no. 1 (1961): 10.

74. Earle, "Myth," 205–8. On the history of guano and its use, see, Stoll, *Larding the Lean Earth*, 187–94; Richard A. Wines, *Fertilizer in America: From Waste Recycling to Resource Exploitation* (Philadelphia: Temple University Press, 1985), 33–70; Edward D. Melillo, "The First Green Revolution: Debt Peonage and the Making of the Nitrogen Fertilizer Trade, 1840–1930," *American Historical Review* 117, no. 4 (October 2012): 1028–60; Genovese, "Cotton, Slavery, and Soil Exhaustion," 8–9.

75. Earle, "Myth," 208
76. Ibid., 210.
77. Gavin Wright, *Old South, New South: Revolutions in the Southern Economy since the Civil War* (Baton Rouge: Louisiana State University Press, 1986), 17. Gray, *History of Agriculture*, 1:450–51. See also Steinberg, *Down to Earth*, 86; Stoll, *Larding the Lean Earth*, 34–36, 132–35.
78. Kirby, *Mockingbird Song*, 87.
79. Stoll, *Larding the Lean Earth*, 131–35. The Stewart County example comes from Eugene Genovese, *The Political Economy of Slavery: Studies in the Economy and Society of the Slave South* (New York: Vintage Books, 1965), 58.
80. Craven, *Soil Exhaustion*, 37.
81. Du Bois, *The Souls of Black Folk*, 52.
82. Bagley, *Soil Exhaustion*, 54.
83. Genovese, *Political Economy of Slavery*, 26, 43.
84. Ibid., 74. Genovese followed in the footsteps of Kenneth Stampp, whose *The Peculiar Institution*, which appeared in 1956 as the civil rights movement was building strength, was a forceful repudiation of the work of U. B. Phillips and others who had suggested that slavery was a benign institution. See Stampp, *The Peculiar Institution: Slavery in the Ante-Bellum South* (New York: Knopf, 1956).
85. Genovese, *Political Economy of Slavery*, 244.
86. Ibid., 55. See also Bagley, *Soil Exhaustion*, 58–65.
87. John Taylor, *Arator: Being a Series of Agricultural Essays, Practical and Political*, 2nd ed. (Washington, D.C.: J. M. Carter, 1814), 69. Jack Temple Kirby also uses Taylor to make this point in *Mockingbird Song*, 81–82. Lewis Gray also connected the overseer system to land abuse. See Gray, *History of Agriculture*, 1:546.
88. Genovese, *Political Economy of Slavery*, 26.
89. Trimble, *Man-Induced Soil Erosion*, 58.
90. Earle, "Myth," 177.
91. I want to acknowledge an unpublished essay by the late Jason Manthorne, "Examining the Southern 'Soil Miner,'" which made the vital case that we need more work on yeomen and their impacts on the region's soils. Copy in my possession.
92. Judith A. Carney, *Black Rice: The African Origins of Rice Cultivation in the Americas* (New York: Cambridge University Press, 2001). Carney's work built on the scholarship of Peter Wood, *Black Majority: Negroes in Colonial South Carolina from 1670 through the Stono Rebellion* (New York, Knopf: 1974); and Daniel Littlefield, *Rice and Slaves: Ethnicity and the Slave Trade in Colonial South Carolina* (Baton Rouge: Louisiana State University Press, 1981). The most influential rebuttal of this thesis came in David Eltis, Philip Morgan, and David Richardson, "Agency and Diaspora in Atlantic History: Reassessing the African Contribution to Rice Cultivation in the Americas," *American Historical Review* 112, no. 5 (2007): 1329–58. See also the "*AHR* Exchange: The Question of 'Black Rice,'" *American Historical Review* 115, no. 1 (February 2010): 123–71. For books that look at slave knowledge in other contexts, see Mart A. Stewart, *What Nature Suffers to Groe: Life, Labor, and Landscape on the Georgia Coast, 1680-1920* (Athens: University of Georgia Press, 1996); Sharla Fett, *Working Cures: Healing, Health, and Power on Southern Slave Plantations* (Chapel Hill: University of North Carolina Press, 2002); Walter Johnson, *River of Dark Dreams: Slavery*

and Empire in the Cotton Kingdom (Cambridge: Harvard University Press, 2013); Dianne D. Glave and Mark Stoll, eds., *To Love the Wind and Rain: African Americans and Environmental History* (Pittsburgh: University of Pittsburgh Press, 2006).

93. Lorena Walsh, *From Calabar to Carter's Grove: The History of a Virginia Slave Community* (Charlottesville: University Press of Virginia, 1997), 63–65, quotes on 63.

94. Taylor, *Arator*, 162–63.

95. Ruffin quoted in Douglas Helms, "Soils and Southern History," *Agricultural History* 74, no. 4 (2000): 749.

96. Moore, *Agriculture in Antebellum Mississippi*, 124.

97. Judith Carney suggested this speculation about the cowpea in personal e-mail correspondence. See also Judith Carney and Richard Nicholas Rosomoff, *In the Shadow of Slavery: African's Botanical Legacy in the Atlantic World* (Berkeley: University of California Press, 2009), 149, 125; Douglas Helms, "Soils and Southern History," *Agricultural History* 74, no. 4 (2000): 749.

98. Bagley, *Soil Exhaustion*, 78.

99. Ibid., 81. More recently, Adam Wesley Dean has made a similar argument in "An Agrarian Republic: How Conflict over Land Use Shaped the Civil War and Reconstruction" (PhD diss., University of Virginia, 2010).

100. Rupert Bayless Vance, *The Human Geography of the South: A Study in Regional Resources and Human Adequacy* (Chapel Hill: University of North Carolina Press, 1932), 74.

101. This is the central argument of Majewski, *Modernizing a Slave Economy*. See also Sarah T. Phillips, "Antebellum Agricultural Reform: Republican Ideology, Sectional Tension," *Agricultural History* 74 (2000): 799–822.

102. These stats are from Arthur Raper, "Gullies and What They Mean," *Social Forces* 16, no. 2 (December 1937): 202–3.

103. Kirby, *Mockingbird Song*, 102.

104. National Emergency Council, *Report on the Economic Conditions of the South*, (Washington, D.C.: GPO, 1938), quoted in David L. Carlton and Peter A. Coclanis, eds., *Confronting Southern Poverty in the Great Depression: The Report on the Economic Conditions of the South with Related Documents* (New York: Bedford-St. Martin's 1996), 48.

105. On Carver's efforts, see Mark Hersey, *My Work Is That of Conservation: An Environmental Biography of George Washington Carver* (Athens: University of Georgia Press, 2011).

106. Charles Johnson, Edwin R. Embree, and W. W. Alexander, *The Collapse of Cotton Tenancy: Summary of Field Studies and Statistical Surveys, 1933–35* (Chapel Hill: University of North Carolina Press, 1935), 20.

107. Kirby, *Mockingbird Song*, 104; Forrest McDonald and Grady McWhiney, "The South from Self-Sufficiency to Peonage: An Interpretation," *American Historical Review* 85, no. 5 (December 1980): 1095–1118; Steinberg, *Down to Earth*, 99–101.

108. C. Vann Woodward, *Origins of the New South, 1877–1913* (Baton Rouge: Louisiana State University Press, 1951), 185.

109. Range, *Century of Georgia Agriculture*, 101.

110. Richard C. Sheridan, "Chemical Fertilizers in Southern Agriculture," *Agricultural History* 53, no. 1 (January 1979): 308; Richard A. Wines, *Fertilizer in America:*

From Waste Recycling to Resource Exploitation (Philadelphia: Temple University Press, 1985), 157–59.

111. Charles Aiken, *Cotton Plantation South since the Civil War* (Baltimore: Johns Hopkins University Press, 1998), 57.

112. Hall, "Soil Erosion," 230; Range, *Century of Georgia Agriculture*, 101–2; Hahn, *Roots of Southern Populism*, 47; Timothy Johnson, "Reconstructing the Soil: Emancipation and the Roots of Chemical-Dependent Agriculture in America," in *The Blue, the Gray, and the Green: Toward an Environmental History of the Civil War*, ed. Brian Allen Drake (Athens: University of Georgia Press, 2015), 191–208.

113. Range, *Century of Georgia Agriculture*, 116.

114. Hahn, "Hunting, Fishing, and Foraging"; Scott Giltner, *Hunting and Fishing in the New South: Black Labor and White Leisure after the Civil War* (Baltimore: Johns Hopkins University Press, 2008).

115. Range, *Century of Georgia Agriculture*, 94, 101, 122; Helms, "Soils and Southern History," 735; Johnson, "Reconstructing the Soil."

116. Steinberg, *Down to Earth*, 245.

117. Aiken, *Cotton Plantation South*, 68–76, quote on p. 68. See also Kirby, *Mockingbird Song*, 105–6; Range, *Century of Georgia Agriculture*, 152.

118. Raper, *Tenants of the Almighty*, 152–53.

119. James Giesen, *Boll Weevil Blues: Cotton, Myth, and Power in the American South* (Chicago: University of Chicago Press, 2011); Range, *Century of Georgia Agriculture*, 268.

120. Raper, *Tenants of the Almighty*, 156–57.

121. Arthur Raper, "Gullies and What They Mean," *Social Forces* 16, no. 2 (December 1937): 201–7.

122. Bennett, *Soil Conservation*, 899.

123. There is a rich new body of literature on cotton, slavery, and plantation agriculture that emphasizes the capitalist dimensions of southern plantation agriculture as well as the importance of cotton and slaves to the birth of both American and global capitalism. See Edward Baptiste, *The Half Has Never Been Told: Slavery and the Making of American Capitalism* (New York: Basic Books, 2014); Sven Beckert, *Empire of Cotton: A Global History* (New York: Knopf, 2014); Walter Johnson, *River of Dark Dreams: Slavery and Empire in the Cotton Kingdom* (Cambridge: Harvard University Press, 2013); Joshua Rothman, *Flush Times and Fever Dreams: A Story of Capitalism and Slavery in the Age of Jackson* (Athens: University of Georgia Press, 2012). For an evolutionary interpretation of cotton and its importance to Anglo-American industrialization, see Edmund Russell, *Evolutionary History: Uniting History and Biology to Understand Life on Earth* (New York: Cambridge University Press, 2011), 103–31.

CHAPTER SEVEN. Somewhere between the Grand Canyon and a Sickening Void

1. Ulrich Bonnell Phillips, *Life and Labor in the Old South* (Boston: Little, Brown, 1929), 3.

2. Mart Stewart has been a leading figure in making this refutation central to the field. See Mart A. Stewart, *What Nature Suffers to Groe: Life, Labor, and Landscape on*

the Georgia Coast, 1680–1920 (Athens: University of Georgia Press, 1996), 5–9; Mart A. Stewart "'Let Us Begin with the Weather?': Climate, Race, and Cultural Distinctiveness in the American South," in *Nature and Society in Historical Context*, ed. Mikulas Teich, Roy Porter, and Bo Gustaffson (New York: Cambridge University Press, 1997), 240–56.

3. Julius Rubin, "The Limits of Agricultural Progress in the Nineteenth-Century South," *Agricultural History* 49 (April 1975): 362–73. His first line is: "I think I had better begin with an apology—or at least an explanation—for the way I approach the subject of the limits of agricultural progress in the nineteenth-century South."

4. Daniel Richter and Daniel Markewitz, *Understanding Soil Change: Soil Sustainability Over Millennia, Centuries, and Decades* (New York: Cambridge University Press, 2001), 40–42, 67. See also Martin Melosi, "Environment," in *Environment*, volume 8 of *The New Encyclopedia of Southern Culture* (Chapel Hill: University of North Carolina Press, 2007), 317; John Majewski, *Modernizing a Slave Economy: The Economic Vision of the Confederate Nation* (Chapel Hill: University of North Carolina Press, 2009), 32–35, 163–65; Douglas Helms, "Soils and Southern History," *Agricultural History* 74, no. 4 (2000): 723–58.

5. I have relied mostly on Jack Temple Kirby's rendition in *Mockingbird Song: Ecological Landscapes of the South* (Chapel Hill: University of North Carolina Press, 2006), 84–92. See also Jack Temple Kirby, *Poquosin: A Study of Rural Landscape and Society* (Chapel Hill: University of North Carolina Press, 1995), 69–74; Jack Temple Kirby, introduction to *Nature's Management: Writings on Landscape and Reform, 1822–1859*, by Edmund Ruffin, ed. Jack Temple Kirby (Athens: University of Georgia Press, 2000): xi–xxxi; Avery Odelle Craven, *Soil Exhaustion as a Factor in the Agricultural History of Virginia and Maryland, 1606–1860* (1926; Gloucester, Mass.: Peter Smith, 1965), 134–43; Steven Stoll, *Larding the Lean Earth: Soil and Society in Nineteenth-Century America* (New York: Hill & Wang, 2002), 150–60; William Mathew, *Edmund Ruffin and the Crisis of Slavery in the Old South* (Athens: University of Georgia Press, 1988).

6. Richter and Markewitz, *Understanding Soil Change*, 118–19; Kirby, *Poquosin*, 112–13; Majewski, *Modernizing a Slave Economy*, 26–35.

7. Helms, "Soils and Southern History," 729–30.

8. Rubin, "Limits of Agricultural Progress," 365.

9. Rupert Bayless Vance, *The Human Geography of the South: A Study in Regional Resources and Human Adequacy* (Chapel Hill: University of North Carolina Press, 1932), 154.

10. Willard Range, *A Century of Georgia Agriculture, 1850–1950* (Athens: University of Georgia Press, 1954), 23–25.

11. C. W. Howard, *A Manual of the Cultivation of the Grasses and Forage Plants at the South*, 2nd ed. (Charleston, S.C.: Walker, Evans & Cogswell, 1875).

12. George Vasey, *Grasses of the South: A Report on Certain Grasses and Forage Plants for Cultivation in the South and Southwest*, U.S. Department of Agriculture, Botanical Division, Bulletin no. 3 (Washington, D.C.: GPO, 1887).

13. Howard, *Manual of the Cultivation*, 1.

14. A. R. Hall, *The Story of Soil Conservation in the South Carolina Piedmont, 1800–1860*, USDA Miscellaneous Publication no. 407 (Washington, D.C.: GPO, 1940), 8.

15. Vance, *Human Geography of the South*, 154–59, quotation on 158.

16. Range, *Century of Georgia Agriculture*, 197, 206. See also Albert G. Way, "'A

Cosmopolitan Weed of the World': Following Bermudagrass," *Agricultural History* 88, no. 3 (Summer 2014): 354–67.

17. Claire Strom, "Texas Fever and the Dispossession of the Southern Yeoman Farmer," *Journal of Southern History* 66, no. 1 (February 2000): 49–74; Paul S. Sutter and Christopher Manganiello, eds., *Environmental History and the American South: A Reader* (Athens: University of Georgia Press, 2009), quote on p. 221; Claire Strom, *Making Catfish Bait out of Government Boys: The Fight against Cattle Ticks and the Transformation of the Yeoman South* (Athens: University of Georgia Press, 2009); Tamara Haygood, "Cows, Ticks, and Disease: A Medical Interpretation of the Southern Cattle Industry," *Journal of Southern History* 52, no. 4 (November 1986): 551–64.

18. Range, *Century of Georgia Agriculture*, 98.

19. Strom, "Texas Fever"; Strom, *Making Catfish Bait*; Rubin, "Limits of Agricultural Progress," 366–68; Majewski, *Modernizing a Slave Economy*, 35–36.

20. Majewski, *Modernizing a Slave Economy*, 16.

21. Ibid., 36, 164–65; Helms, "Soils and Southern History," 732–33.

22. Stoll, *Larding the Lean Earth*, 127.

23. I make this argument in more detail in my foreword to Strom, *Making Catfish Bait*, xiii–xvi.

24. Vance, *Human Geography of the South*, 103.

25. Stanley Trimble, *Historical Agriculture and Soil Erosion in the Upper Mississippi Valley Hill Country* (Boca Raton, Fla.: CRC Press, 2013), xlv–xlvi.

26. On rainfall erosivity, see Stoll, *Larding the Lean Earth*, 139–40; Stanley W. Trimble, *Man-Induced Soil Erosion on the Southern Piedmont, 1700–1970*, enhanced ed. (Ankeny, Iowa: Soil and Water Conservation Society, 2008), 12–13; Francis J. Magilligan and Melissa L. Stamp, "Historical Land-Cover Changes and Hydrographic Adjustment in a Small Georgia Watershed," *Annals of the American Association of Geographers* 7, no. 4 (1997): 619.

27. This point is adapted from one that Stanley Trimble makes in *Man-Induced Soil Erosion*, 12.

28. Arthur R. Hall, "Soil Erosion and Agriculture in the Southern Piedmont: A History" (PhD diss., Duke University, 1948), 3.

29. Stanley Trimble, "The Historical Decrease of Soil Erosion in the Eastern United States—The Role of Geography and Engineering," in *Engineering the Earth: The Impacts of Megaengineering Projects*, ed. Stanley Brunn (New York: Springer, 2011), 1383–93. In personal correspondence, Dan Richter has emphasized the importance of land clearance to diminished evapotranspiration and increased surface runoff.

30. David R. Montgomery, *Dirt: The Erosion of Civilizations* (Berkeley: University of California Press, 2007), 20.

31. Robert W. McVety, "Steep-Sided Gully Erosion in Stewart County, Georgia: Causes and Consequences" (master's thesis, Florida State University, 1971), 13, 23; Hall, "Soil Erosion," 2–9; Trimble, *Man-Induced Soil Erosion*, 4.

32. Trimble, *Man-Induced Soil Erosion*, vii.

33. William W. Winn, "The View from Dowdell's Knob," in *The New Georgia Guide* (Athens: University of Georgia Press, 1996), 366–67; Helen Elisa Terrill, *History of Stewart County, Georgia*, ed. Sara Robertson Dixon (Columbus, Ga.: Columbus Office Supply, 1958), 1:72–73.

34. Majewski, *Modernizing a Slave Economy*, 24; census figures from Historical

Census Browser, University of Virginia, Geospatial and Statistical Data Center, http://mapserver.lib.virginia.edu/, accessed October 25, 2014.

35. On terracing, see McVety, "Steep-Sided Gully Erosion," 43, 75.

36. David C. Long, et al., "Soil Survey of Stewart County, Georgia," in *Field Operations of the Bureau of Soils, 1913* (Washington, D.C.: GPO, 1914), 18.

37. Winn, "View from Dowdell's Knob," 376–78.

38. Stats from McVety, "Steep-Sided Gully Erosion," 46.

39. Historical Census Browser.

40. U.S. Census Bureau, State and County QuickFacts, Stewart County, Georgia, http://quickfacts.census.gov/qfd/states/13/13259.html, accessed July 31, 2013.

41. Long et al., "Soil Survey of Stewart County," 19.

42. Lisa G. Joyce, *Geologic Guide to Providence Canyon State Park*, Geologic Guide 9 (Atlanta: Georgia Geological Survey, 1985); McVety, "Steep-Sided Gully Erosion," 30–32.

43. The information from this paragraph is taken from Joyce, *Geologic Guide*, 1–12.

44. McVety, "Steep-Sided Gully Erosion," 66–68.

45. Joyce, *Geologic Guide*, 10–12, McVety, "Steep-Sided Gully Erosion," 68–69; D. Hoye Eargle, *Stratigraphy of the Outcropping Cretaceous Rocks of Georgia*, U.S. Geological Survey Bulletin no. 1014 (Washington, D.C.: GPO, 1955), 59–60, 71–72.

46. Joyce, *Geologic Guide*, 10–12, McVety, "Steep-Sided Gully Erosion," 68–69. On kaolin, see Charles Seabrook and Marcy Louza, *Red Clay, Pink Cadillacs, and White Gold: The Kaolin Chalk Wars* (Atlanta: Longstreet Press, 1995).

47. "Providence Canyons Erode Rapidly as Sands Wash Away," *Stewart-Webster Journal*, July 7, 1938, 3.

48. H. Andrew Ireland, "'Lyell Gully, a Record of a Century of Erosion," *Journal of Geology* 47, no. 1 (January–February 1939): 61.

49. A. R. Hall, *Early Erosion Control Practices in Virginia*, USDA Miscellaneous Publication 256 (Washington, D.C.: GPO, 1937), 10. See also H. A. Ireland, C. F. S. Sharpe, and D. H. Eargle, *Principles of Gully Erosion in the Piedmont of South Carolina*, USDA Technical Bulletin no. 633 (Washington, D.C.: GPO: 1939), 40–42.

50. Hall, "Soil Erosion," 34.

51. See Trimble, *Man-Induced Soil Erosion*, 8.

52. Vance, *Human Geography of the South*, 87; Helms, "Soils and Southern History," 746–47, 749–51.

53. Neil Maher uses the Coon Creek watershed as a case study in his book *Nature's New Deal: The Civilian Conservation Corps and the Roots of the American Environmental Movement* (New York: Oxford University Press, 2008), quote on p. 122. See also Trimble, *Historical Agriculture*, 98; Lynne Heasley, *A Thousand Pieces of Paradise: Landscape and Property in the Kickapoo Valley* (Madison: University of Wisconsin Press, 2005), 21–46.

54. This paragraph is drawn from Marsha Weisiger, *Dreaming of Sheep in Navajo Country* (Seattle: University of Washington Press, 2009), 31–60. See also Richard White, *The Roots of Dependency: Subsistence, Environment, and Social Change among the Choctaws, Pawnees, and Navajos* (Lincoln: University of Nebraska Press, 1983), 212–314.

55. Kate Showers, "Soil Erosion in the Kingdom of Lesotho: Origins and Colonial Response, 1830s–1950s," *Journal of Southern African Studies* 15, no. 2 (1989): 263–86; Kate B. Showers, *Imperial Gullies: Soil Erosion and Conservation in Lesotho* (Athens: Ohio University Press, 2005); James McCann, *Green Land, Brown Land, Black Land:*

An Environmental History of Africa, 1800–1990 (Portsmouth, N.H.: Heinemann, 1999), 141–73.

EPILOGUE. The Ecology of Erasure

1. Leslie Kimel, "Georgia Wildlife Federation," New Georgia Encyclopedia, August 20, 2013, http://www.georgiaencyclopedia.org/articles/geography-environment/georgia-wildlife-federation, accessed October 11, 2013.

2. Ann Vileisis, *Discovering the Unknown Landscape: A History of America's Wetlands* (Washington, D.C.: Island Press, 1997), 244–52.

3. Ibid., quote on p. 249.

4. On these postwar developments, see Paul Hirt, *A Conspiracy of Optimism: Management of the National Forests since World War Two* (Lincoln: University of Nebraska Press, 1994); Mark W. T. Harvey, *A Symbol of Wilderness: Echo Park and the American Conservation Movement* (Albuquerque: University of New Mexico Press, 1994); Kevin Marsh, *Drawing Lines in the Forest: Creating Wilderness in the Pacific Northwest* (Seattle: University of Washington Press, 2007); Edmund Russell, *War and Nature: Fighting Human and Insect Enemies from World War I to Silent Spring* (New York: Cambridge University Press, 2001); Joshua Blu Buhs, *The Fire Ant Wars: Nature, Science, and Public Policy in Twentieth-Century America* (Chicago: University of Chicago Press, 2004); Pete Daniel, *Toxic Drift: Pesticides and Health in the Post–World War Two South* (Baton Rouge: Louisiana State University Press, 2005).

5. Vileisis, *Discovering the Unknown Landscape*, 245–46. See also Ben B. Blackburn and George Laycock, "Where Conservation Is a Bad Word," *Field & Stream*, December 1969, 12–14, 58–62.

6. Vileisis, *Discovering the Unknown Landscape*, 245–46; Charles H. Wharton, *The Southern River Swamp: A Multiple-Use Environment* (Atlanta: Georgia State University, 1970). On Earth Day, see Adam Rome, *The Genius of Earth Day: How a 1970s Teach-in Unexpectedly Made the First Green Generation* (New York: Hill & Wang, 2013).

7. On Wharton and his importance, see Vileisis, *Discovering the Unknown Landscape*, 246; H. Harold Brown, *The Greening of Georgia: The Improvement of the Environment in the Twentieth Century* (Macon: Mercer University Press, 2002), 44; Wharton, *Southern River Swamp*, i, 1. See also John and Mildred Teal, *Life and Death of the Salt Marsh* (Boston: Little, Brown, 1969). John Teal wrote an early and influential study on salt marsh ecology that took Georgia salt marshes as its subject: John M. Teal, "Energy Flow in the Salt Marsh Ecosystem of Georgia," *Ecology* 43, no. 4 (October 1962): 614–24. On Odum's influence, see Vileisis, *Discovering the Unknown Landscape*, 217–18.

8. Wharton, *Southern River Swamp*, 1.

9. Stanley W. Trimble, "Culturally Accelerated Sedimentation on the Middle Georgia Piedmont," (master's thesis, University of Georgia, 1969), 4.

10. Stanley W. Trimble, "The Alcovy River Swamps: The Result of Culturally Accelerated Sedimentation," *Bulletin of the Georgia Academy of Science* 28 (September 1970): 131–41.

11. Ibid.

12. Ibid., 131–37; Montgomery quoted on p. 137. See also H. H Barrows, J. V. Phillips, and J. E. Brantly, *Agricultural Drainage in Georgia*, Geological Survey of Georgia Bulletin no. 32 (Atlanta: Byrd Printing, 1917), x.

13. Trimble, "Alcovy River Swamps," 139.

14. Ibid., 138.

15. Albert C. Staheli, David E. Ogren, and Charles H. Wharton, "Age of Swamps in the Alcovy River Drainage Basin," *Southeastern Geology* 16, no. 2 (1974): 103–6; Stanley W. Trimble, "The Age of the Alcovy River Swamps: A Discussion," *Southeastern Geology* 18, no. 3 (1977): 191–94; Albert C. Staheli, David E. Ogren, and Charles H. Wharton, "Age of Swamps in the Alcovy River Drainage Basin: A Reply," *Southeastern Geology* 18, no. 3 (1977): 195–98.

16. Leon Sisk, *The Changed Look of the Countryside* (Franklin Springs, Ga.: Advocate Press, 1975), 127–28.

17. Will Anderson, "A Small, Important River: Alcovy's Ecosystem Attracts Scientists and Preservationists," *Atlanta Journal-Constitution*, January 3, 1997. On the possible effects of upland soil erosion on salt marshes, see John Paul Schmidt and Douglas Patton, "The Value of Ecosystem Services Provided by Georgia's Coastal Marshes," unpublished report to the Georgia Department of Natural Resources, April 2014, 9.

18. Albert E. Cowdrey, *This Land, This South: An Environmental History*, rev. ed. (Lexington: University Press of Kentucky, 1996), 127.

19. My thinking here has been influenced by William Cronon, "The Riddle of the Apostle Islands," *Orion*, May–June 2003, 36–42. See also James Feldman, *A Storied Wilderness: Rewilding the Apostle Islands* (Seattle: University of Washington Press, 2011).

20. Thomas D. Clark, *The Greening of the South: The Recovery of Land and Forest* (Lexington: University Press of Kentucky, 1984), 146.

21. This phrase is borrowed from Bill McKibben, "An Explosion of Green," *Atlantic*, April 1995, http://www.theatlantic.com/magazine/archive/1995/04/an-explosion-of-green/305864, accessed October 29, 2014.

22. William Boyd, "The Forest Is the Future? Industrial Forestry and the South Pulp and Paper Complex," in *The Second Wave: Southern Industrialization, 1940–1970s*, ed. Philip Scranton (Athens: University of Georgia Press, 2001), 168–218; Michael Williams, *Americans and Their Forests: A Historical Geography* (New York: Cambridge University Press, 1989), 238–88.

23. See Sara Gregg, *Managing the Mountains: Land Use Planning, the New Deal, and the Creation of a Federal Landscape* (New Haven: Yale University Press, 2010); Darwin Lambert, *The Undying Past of Shenandoah National Park* (Boulder, Colo.: Robert Rinehart, 1989); Daniel S. Pierce, *The Great Smokies: From Natural Habitat to National Park* (Knoxville: University of Tennessee Press, 2000); Durwood Dunn, *Cades Cover: The Life and Death of a Southern Appalachian Community* (Knoxville: University of Tennessee Press, 1988); Kathryn Newfont, *Blue Ridge Commons: Environmental Activism and Forest History in Western North Carolina* (Athens: University of Georgia Press, 2012); Margaret Lynn Brown, *Wild East: A Biography of the Great Smoky Mountains* (Gainesville: University Press of Florida, 2001).

24. See Albert G. Way, *Conserving Southern Longleaf: Herbert Stoddard and the Rise of Ecological Land Management* (Athens: University of Georgia Press, 2011); William Boyd, *New South, New Nature: Regional Industrialization and Environmental Change in the Post–New Deal American South* (Baltimore: Johns Hopkins University Press, 2015).

25. Germaine M. Reed, "Charles Herty (1867–1938)," New Georgia Encyclopedia, August 19, 2013, http://www.georgiaencyclopedia.org/articles/business-economy/charles-herty-1867-1938, accessed October 29, 2014.

26. Jonathan Daniels, *The Forest Is the Future* (New York: International Paper Company, 1957), quote on p. 7, acreage figure on p. 38.

27. Boyd, "Forest Is the Future?" On kudzu, see Derek H. Alderman and Donna D'Segner Alderman, "Kudzu: A Tale of Two Vines," *Southern Cultures* 7, no. 3 (Fall 2001): 49–64.

28. The discussion of the Soil Bank program is based on Boyd, "Forest Is the Future?," 191–93.

29. On the creation and early history of Sumter National Forest, see Percy J. Paxton, "The National Forests and Purchase Units of Region Eight," July 1, 1950, National Archives and Records Administration, Southeast Region, Land Acquisition and Historical Files, box 1, folder 1 ("Land Classification-General-Historical"), 51–52, record group 95; Tara Mitchell Mielnik, *New Deal, New Landscape: The Civilian Conservation Corps and South Carolina's State Parks* (Columbia: University of South Carolina Press, 2011), 72; "History of the Enoree and Long Cane Ranger Districts," U.S. Forest Service, n.d., http://www.fs.usda.gov/detail/scnfs/learning/history-culture/?cid=fsbdev3_037408, accessed November 20, 2013; Daniel Richter and Daniel Markewitz, *Understanding Soil Change: Soil Sustainability over Millennia, Centuries, and Decades* (New York: Cambridge University Press, 2001), 58–63. On the history of the national forests in the South, see Gerald W. Williams, "Private Property to Public Property: The Beginnings of the National Forests in the South," paper presented at the American Society for Environmental History Meeting, Providence, Rhode Island, March 2003, copy in author's possession.

30. Richter and Markewitz, *Understanding Soil Change*, 51.

31. Louis J. Metz, *The Calhoun Experimental Forest* (n.p.: n.p., 1958), 1, http://www.srs.fs.usda.gov/pubs/misc/metz_calhoun.pdf, accessed November 20, 2013.

32. National Forest Reservation Commission, "The Report of the National Forest Reservation Commission for the Year Ended June 30, 1930," 71st Cong., 3rd sess., Senate Document no. 223 (Washington, D.C.: GPO, 1930), 18–19.

33. Christopher Johnson and David Govatski, *Forests for the People: The Story of America's Eastern National Forests* (Washington, D.C.: Island Press, 2013), 145–66.

34. H. H Wooten, *The Land Utilization Program, 1934–1946: Origin, Development, and Present Status*, USDA Agricultural Economic Report no. 185 (Washington, D.C.: GPO, 1965), v–vi.

35. Robert G. Pasquill Jr., *Planting Hope on Worn-Out Land: The History of the Tuskegee Land Utilization Project, Macon County Alabama, 1935–1959* (Montgomery, Ala.: New South Books, 2008), 17.

36. Quoted in Sarah T. Warren and Robert E. Zabawa, "The Origins of the Tuskegee National Forest: Nineteenth- and Twentieth-Century Resettlement and Land Development Programs in the Black Belt Region of Alabama," *Agricultural History* 72, no. 2 (Spring 1998): 498.

37. Pasquill, *Planting Hope on Worn-Out Land*, 37.

38. Ibid., 19–40.

39. Ibid., 71–126; Warren and Zabawa, "Origins of the Tuskegee National Forest"; Dwight D. Eisenhower, "Proclamation 3326—'Establishing the Tuskegee National Forest, Alabama, the Oconee National Forest, Georgia, and the Tombigbee National Forest. Mississippi," November 27, 1959, American Presidency Project, http://www.presidency.ucsb.edu/ws/?pid=107425, accessed January 21, 2015. Another of the state's

LUPS, the West Alabama Planned Development Project, dissolved in 1938 after turning over to the U.S. Forest Service almost ninety thousand acres that became part of the Talladega National Forest.

40. Wooten, *Land Utilization Program*, 41–42; Raper, *Tenants of the Almighty* (New York: MacMillan Company, 1943), 195; "Georgia's New National Forest Areas," n.d. (but likely 1959 or 1960), National Archives and Records Administration, Southeast Region, Land Acquisition and Historical Files, box 1, Folder 5500 ("Land Classification, Georgia, Historical"), record group 95.

41. "Georgia's New National Forest Areas."

42. For the Boswell story, see Mark Arax and Rick Wartzman, *The King of California: J. G. Boswell and the Making of a Secret American Empire* (New York: Public Affairs, 2003); Jerry Hirsch, "James G. Boswell II Dies at 86; Cotton Magnate Built Family Farm into Agribusiness Giant," *Los Angeles Times*, April 7, 2009.

43. Mary Summers, "The New Deal Farm Programs: Looking for Reconstruction in American Agriculture," *Agricultural History* 74, no. 2 (2000): 243.

44. Raper, *Tenants of the Almighty*, 203–9.

45. Ibid., 220–21.

46. Wooten, *Land Utilization Program*, 41–42; "Georgia's New National Forest Areas."

47. Wooten, *Land Utilization Program*, 45–55; R. Douglas Hurt, "The National Grasslands: Origin and Development in the Dust Bowl," *Agricultural History* 59, no. 2 (1985): 246–59; U.S. Forest Service National Grasslands website, http://www.fs.fed.us/grasslands/, accessed October 31, 2014. Geoff Cunfer argues that, despite the rhetoric of program officials, little of this land had actually been in cultivation. See Cunfer, "The New Deal's Land Utilization Program in the Great Plains," *Great Plains Quarterly* 21 (Summer 2001): 193–210.

48. Andrew Isenberg, *Mining California: An Ecological History* (New York: Hill & Wang, 2005), 23–51, 165–76.

INDEX

Agee, James, 184; *Let Us Now Praise Famous Men* (with Evans), 4–6
Agricultural Adjustment Act (AAA), 197
Albany, Ga., plantation region, 14
Alcovy Conservation Center, 185, 189–91
Alcovy River swamps, 185–93
Alexander, Will W., 152
Alfisols, 159, 165
American Oasis, The (Higbee), 106–7
Anson County, N.C., 49, 127
Anti-Debris Association, 205
Antiquities Act (1906), 36
Appleman, Roy E., 75
Arator (Taylor), 146
Arnall, Ellis, 1; *The Shore Dimly Seen*, 97–98
Atlanta Constitution, 65, 66, 67, 78

babesiosis (Texas fever), 164–65
Bagley, William Chandler, 149–50
Bailey, Liberty Hyde, 43
Battle of Shepherd's Plantation, 19
Beck, M. W. 37
Behold This Land (R. Lord), 92, 103
Bennett, Hugh Hammond, 37–38, 111–12, 114, 127, 155, 156; Dust Bowl and, 89; early work with National Cooperative Soil Survey, 49–51; friend of Jonathan Daniels, 95–96; Friends of the Land and, 91; as head of Soil Conservation Service, 92; invocation of Stewart County, 58, 89–90, 105; Navajo Reservation erosion and, 181; New Deal conservation and, 84; 1920s international experience, 59–60; Progressive conservation and, 52–54, 56; southern soil surveys and, 54–60

—works: *Elements of Soil Conservation*, 90; *Soil Conservation*, 90; *Soil Erosion, a National Menace* (with Chapline), 85; *Soils and Agriculture of the Southern States*, 58; *Soils of Cuba*, 59
Birth of a Nation, The (Griffith), 45
"Black Sunday," 89
boll weevil, 154–55
book farming, 21, 126
Boswell, James G., 203
Brasstown Bald, 76
Brink, Wellington, 59, 95
Brooks, Paul, 5
Bryce Canyon, 4, 73, 75
Buchanan, James, 85
Burch, B. F., 75
Burdette, W. W., 37
Bureau of American Ethnology, 52
Bureau of Soils. *See* U.S. Bureau of Soils
Burton, Glenn, 164

Caldwell, Erskine, 76
Calhoun Critical Zone Observatory, 198–99
California Gold Rush, 25
Canyons at Providence, The (Flanagan), 78–81
Carr, Earl, 55
Carter, Hugh, 81
Carter, Jimmy, 3, 13, 81
Carver, George Washington, 49, 152
catastrophism, 27
Chamberlain, Thomas Nelson, 52
Chapline, W. R., 85
Chase, Stuart, 105, 176; *A New Deal*, 98; *Rich Land, Poor Land*, 98–102
Chattahoochee Valley Chamber of Commerce, 71

Chincha Islands, 140–41
civil rights movement, 106
Civil War, 149–50
Civilian Conservation Corps (CCC), 67, 73–74, 75–76, 99, 197, 203–4
Clansman, The (Dixon), 45
Clarke-McNary Act (1924), 197, 198
Clayton Formation, 172–77
Collapse of Cotton Tenancy, The (C. Johnson, Embree, and Alexander), 152
Collier, John, 86, 181
Columbian Exchange, 147
Columbus Chamber of Commerce, 71, 73, 75
Columbus Enquirer, 66, 68–75
Commission on Interracial Cooperation (CIC), 155
Connecticut Agricultural Experiment Station, 40
Connor, A. B., 85
Conservation of Natural Resources, The (Van Hise), 52
Consumers' Research, 98
Consumers Union, 98
contour plowing, 125–26
Cooke, Morris, 91
Coon Creek Demonstration Area, 179
Cooper, James Fenimore, 22
Cotting, John Ruggles, 24–25, 28–29, 31, 34
Country Life Commission, 52
Country Life Movement, 43, 52
cowpeas, 140, 148–49
Craven, Avery Odelle, 139, 144, 145; *Soil Exhaustion as a Factor in the Agricultural History of Virginia and Maryland*, 129
creationism, 107–8
Creek Indians, 14, 16, 17, 19, 130
crop-lien laws, 151–52

Dabney, Charles, 40–41, 49
Daniels, Jonathan W., 105, 135, 177; *The Forest Is the Future*, 196–97; *A Southerner Discovers the South*, 95–97
Daniels, Josephus, 95

Darwin, Charles, 27
Davis, Royal O., 58, 135–36
Davy, Humphrey, 161; *Elements of Agricultural Chemistry*, 38
Delano, Jack, 83, 136
Division of Agricultural Soils, 41. *See also* U.S. Bureau of Soils
Division of Soils, 41. *See also* U.S. Bureau of Soils
Dixie Highway, 66
Dixon, Thomas, 45
Driftless Area, 179–80
Du Bois, W. E. B., 14, 144
Dunbar, William, 126
Dust Bowl, 83–84, 87, 89, 111; mentioned, 37, 63, 105, 181
Dutton, Clarence, 32

Earle, Carville, 136, 139–42, 146, 148, 149
East Alabama Soil Conservation District, 202
Edwards Woodruff v. North Bloomfield Mining and Gravel Company, 205, 207
Eisenhower, Dwight, 202
Elements of Agricultural Chemistry (Davy), 38
Elements of Soil Conservation (Bennett), 90
Elliot, Charles M., 75
Ellis, Mrs. F. W., 72
Embree, Edwin R., 152
Evans, Walker, 83, 116, 184; *Let Us Now Praise Famous Men* (with Agee), 4–6
Exploration of the Colorado River of the West (Powell), 35

F. D. Roosevelt State Park, 74
Fairfield County, S.C., 54–56, 85, 86, 87; mentioned, 59, 112, 179, 198
Fall Line Hills, 13–14, 16, 31, 33
Farm Security Administration (FSA), 83, 103, 116
Farm That Won't Wear Out, The (Hopkins), 44
Farmers of Forty Centuries (F. H. King), 43–45
Fechner, Robert, 73–74

Federal Watershed Protection and Flood Prevention Act (1954), 185
Federal Writers' Project, 76
Ferguson, Bruce, 121–22
Fisher, Carl, 66
Flaherty, Francis, 103
Flaherty, Robert J., 102–5
Flanagan, Thomas Jefferson, 77–82, 156, 192; *The Canyons at Providence*, 78–81; *Jimson Weeds*, 78
floodplain sedimentation, 114–23
Florence Marina and State Park, 20
Florida Short Route, 66, 67
Ford, Henry, 63
Ford, John, 104
Forest Is the Future, The (J. W. Daniels), 196–97
Fort Benning, 13, 29, 72
Fortune magazine, 4–5
Friends of the Land, 91

Garden of the Gods, 73
Geiger, Robert, 89
Genovese, Eugene, 143–50
Geological Survey of Georgia, 11, 24, 26, 28, 31, 33–34; Alcovy River swamps and, 189; Piedmont drainage and, 123
George, Walter, 73–74
Georgia: The WPA Guide to Its Towns and Countryside (WPA), 76–77, 82
Georgia First Association, 66
Georgia Wildlife Federation (GWF), 185, 189–92
Georgia's "Little Grand Canyon," 34, 73, 75, 81, 82; mentioned, 1, 11, 36, 64, 68, 108
Gilbert, Grove Karl, 32
Governors' Conference on Conservation of Natural Resources, 51–53
Grand Canyon, 4, 35–36, 65, 73, 208
Grapes of Wrath, The (Steinbeck), 104
grass husbandry, 162–64
Gray, Lewis Cecil, 129, 143, 144
Greene, Nathanael, 119
Greene County, Ga., 119–21, 136, 155, 179, 202–4
"greening of the South," 195–97

Grierson, John, 103, 104
Griffith, D. W., 45
guano, 140–41

Hale County, Ala., 4, 184
Hall, Arthur R., 118, 125, 126, 178
Happ, Stafford C., 115–18, 199
Hard Labor Creek State Park, 204
Hardwick, Richard Shipp, 126
Harris, Joel Chandler, 202
Hatch Act (1887), 40
Henry, Patrick, 21–22
Herty, Charles, 196
Higbee, Edward: *The American Oasis*, 106–7; *The Squeeze*, 107
highway progressives, 66, 71
Hilgard, Eugene W., 25, 32, 41–43
hillside ditches, 126–27
History of Stewart County, Georgia (Terrill), 70
Hitchiti Experimental Forest, 204
Holly Springs National Forest, 199
Homestead Act, 51
Homochitto National Forest, 199
Hopkins, Cyril, 43–46, 60, 129; *The Farm That Won't Wear Out*, 44; *Soil Fertility and Permanent Agriculture*, 44; *The Story of the Soil*, 44–45
horizontal plowing, 125–26
House Public Lands Committee, 87, 89
Howard, Albert, 45
Howard, Charles Wallace, 163
Howell, B. E., 86
Human Geography of the South, The (Vance), 150
Humber photograph, 22–23
humus theory, 38–39

Imperial Bureau of Soil Science, 94
Ingram and LeGrand Lumber Company, 81
Ireland, H. Andrew, 177–78

Jacks, G. V., 94
Jasper County, Ga., 202, 204
Jefferson, Thomas, 125–26, 134, 149
Jim Crow, 152, 155–56

Jimson Weeds (Flanagan), 78
Johnson, Charles, 152
Johnson, Samuel, 40
Jones County, Ga., 202, 204

kaolin, 176
King, Clarence, 32
King, Franklin Hiram, 41–46, 60, 129; *Farmers of Forty Centuries*, 43–45

Lake Walter F. George, 13, 20
Land, The (Flaherty), 102–5
land utilization projects (LUPs), 200–204
Lange, Dorothea, 83
Lanier, Sidney, 79
Lauderdale County, Miss., 54, 57
Law, Thomas Cassells, 131
Lay of the Last Minstrel, The (Scott), 78
Lee, Russell, 83
Leopold, Aldo, 91
Lesotho, 182–83
Let Us Now Praise Famous Men (Agee and Evans), 4–6
Liebig, Justus von, 38–41; *Organic Chemistry and Its Applications to Agriculture and Physiology*, 38
Life and Death of a Salt Marsh (Teal and Teal), 188
Life and Labor in the Old South (Phillips), 158
Lincoln Highway, 65–66
Lions Club of Lumpkin, 66, 68, 69, 71, 73, 74–75
Little, George, 31
Loess Hills, 85, 115–17, 179, 180
Long, David, 37, 56–57, 172
Lord, Kate, 92
Lord, Russell, 91–93, 103–4; *Behold This Land*, 92, 103; *To Hold This Soil*, 92
Lorentz, Pare, 103–4; *The Plow That Broke the Plains*, 103; *The River*, 103
Louisa County, Va., 50–51, 112
Lowdermilk, Walter, 44, 90
Lowe, David Walker, 19–20
Lowe, Jane Dorsey, 19
Lowndes County, Ala., 114
Lumpkin Independent, 22

Lumpkin Lions Club. *See* Lions Club of Lumpkin
Lyell, Charles, 27–31, 34, 115; *Principles of Geology*, 27, 29; *A Second Visit to the United States*, 28–29
Lyell Gully, 27–31, 34, 96, 177

Macdonald, Dwight, 5
Macon County, Ala., 49, 200–202
Madison, James, 128
Majewski, John, 132, 150, 165–66
Malakoff Diggins State Historic Park, 205–8
malaria, 117–18
Man and Nature (Marsh), 21
Manual of the Cultivation of the Grasses and Forage Plants at the South, A (Howard), 163
Marbut, Curtis F., 48, 58
Marsh, George Perkins, 21
McCall, A. G., 59–60
McCollum, Jerry, 190–91
McGee, W J (William John), 51–53
McVety, Robert, 17, 18
meadow, 55–57, 114, 123
Mississippian cultures, 14, 16
mixed husbandry, 133–34, 140, 143, 164
Morris, F. Grave, 93–94
Muir, John, 51, 207
Mulberry Grove Plantation, 119
Murder Creek watershed, 114, 118, 122, 202

Natchez District, 126, 140, 199
National Conservation Commission, 53
National Cooperative Soil Survey, 11, 12, 37–38, 41, 46–51, 56
National Geographic Society, 52
National Park Service, 75
National Resources Board (NRB), 86–87, 200; *Soil Erosion: A Critical Problem in American Agriculture*, 87
National Wild and Scenic Rivers Act, 188
Navajo Reservation, 87, 181–82
New Deal: congressional hostility to, 104, 105; conservation during, 6–7,

53, 67, 84–86, 102; documentary impulse during, 5–6, 8–9, 102–5; environmental history of, 8; environmental narrative of, 84, 102; and Providence Canyon park campaign, 74; road building and recreational development during, 64
New Deal, A (Chase), 98
North Bloomfield Mining and Gravel Company, 205
North Central Georgia Project, 202–3

Oconee National Forest, 118–19, 121, 202–4
Odum, Eugene, 188, 191
Odum, Howard, 155, 188; *Southern Regions of the United States*, 112
Okefenokee Swamp, 2, 123
Olmsted, Frederick Law, 132
open range, 134–35, 153–54, 162
organic agriculture movement, 45–46
Organic Chemistry and Its Applications to Agriculture and Physiology (Liebig), 38
Origin and Nature of Soils, The (Shaler), 32
Outdoor Recreational Resource Review Commission, 81
overseer system, 145–46

Pace, Steve, 73–74
Panama Canal Zone, 59
Parks, Gordon, 83
Patton, George, 72
permanent agriculture, 43–46, 56, 60, 91, 106, 129
Phillips, U. B., 144; *Life and Labor in the Old South*, 158
phosphate fertilizers, 140–41, 153–54
Piedmont National Wildlife Refuge, 204
Piedmont Plantation Project, 202
Pinchot, Gifford, 51–52
plantation management, 154, 156
Plow that Broke the Plains, The (Lorentz), 103
Poullain, Thomas, 121
Powell, John Wesley, 32, 35, 41, 52; *Exploration of the Colorado River of the West*, 35
Preface to Peasantry (Raper), 155
Preliminary Report on the Geology of the Coastal Plain of Georgia (Veatch and Stephenson), 33
Principles of Geology (Lyell), 27, 29
Progressive conservation movement, 51–54, 56, 58, 84
Providence Canyon: Alcovy River swamps and, 191–93; author's first visit to, 3–4; compared with other areas of extreme gullying, 180–84; Flanagan's interpretation of, 77–81; geologists' encounters with, 24, 29, 33–34, 36; geology of, 172–78; historiographical importance of, 6–9; and Malakoff Diggins, 205–8; nineteenth-century development of, 20, 22; park campaign, 63–77; post–World War II mentions of, 105–8; rethinking interpretation of, 109–11, 119, 123, 127, 147, 157, 170; soil conservationists' interpretation of, 83–84, 90–105; soil surveyors' encounters with, 37, 57–58, 60; southern conservation and, 194–95, 204
Providence Canyon State Park, 1–3, 8, 13, 16, 157, 183; creation of, 81–82; mentioned, 11
Providence "Caves," 57, 67–69, 92–93, 176
Providence Formation, 172–77
Providence Methodist Church, 20, 57, 99
Pulp and Paper Laboratory, 196
Putnam County, Ga., 130, 202

rainfall erosivity, 166–68
Randolph, Thomas Mann, 125
Raper, Arthur, 120, 136, 155–56; *Preface to Peasantry*, 155; *Sharecroppers All*, 155; *Tenants of the Almighty*, 155, 203; *The Tragedy of Lynching*, 155
Reconnaissance Erosion Survey of the United States, 87, 89, 111–12, 200
recyclic shifting cultivation, 139, 141, 148

Report of the Country Life Commission (Country Life Commission), 52
Report of the National Conservation Commission (National Conservation Commission), 53
Report on the Economic Conditions of the South, 152
Resettlement Administration, 83, 103
Rich Land, Poor Land (Chase), 98–102
Ripley Formation, 173, 175
River, The (Lorentz), 103
Rivers, E. D., 75, 97
Roanoke settlement (Ga.), 19
Robinson, Solon, 130
Rock Eagle 4-H Center, 204
Rodale, Jerome, 45
Rogers, William Barton, 25
Roosevelt, Franklin Delano, 63–64, 74, 86, 98–99, 198–99, 204; mentioned, 53, 95, 155, 181
Roosevelt, Theodore, 17, 36, 51, 53, 207
Rothamsted experiments, 40, 44
Rothstein, Arthur, 6, 83, 103
rough gullied land, 17, 55–57
Royal Gorge (Arkansas River), 72–73
Rubin, Julius, 158
Ruffin, Edmund, 25, 114, 143, 148, 159–61
Russell, Richard, 73

Sandy Creek Demonstration Project, 94
saprolite, 169
Sauer, Carl, 93, 127
Sawyer, Lorenzo, 207
Scott, Walter, 78
Scull Shoals, 119–23, 202, 204
Sears, Paul, 91
Second Report on the Clay Deposits of Georgia (Veatch), 33
Second Visit to the United States, A (Lyell), 28–29
sedimentation, 114–23
"See America First," 36, 65
"See Georgia First," 66–67, 76
Shahn, Ben, 83
Shaler, Nathaniel Southgate, 32
Sharecroppers All (Raper), 155
Sherman's March, 119, 189

shifting cultivation, 129–42, 148, 150, 153, 161–62, 165–66; in Stewart County, 171
Shipp, Nelson, 66, 68, 69, 73
Shore Dimly Seen, The (Arnall), 97–98
Singer, Sam, 177
Sisk, Leon, 94
slavery, 142–51, 171
Small Watershed Program, 185–87
Social Forces (journal), 155
Soil and Civilization (M. Whitney), 54
Soil Bank program, 186, 197
soil classification, 46–49
Soil Conservation (Bennett), 90
Soil Conservation (magazine), 94–95
Soil Conservation Act, 89
Soil Conservation and Domestic Allotment Act, 197
Soil Conservation Service (SCS), 67, 94, 106, 127, 197; Bennett and, 37, 89, 92; southern national forests and, 202–4; stream channelization and, 185–91; mentioned, 115, 177
Soil Erosion: A Critical Problem in American Agriculture (National Resources Board), 87
Soil Erosion, a National Menace (Bennett and Chapline), 85
Soil Erosion and Moisture Conservation Investigations, 85
Soil Erosion Service (SES), 86–87
Soil Exhaustion and the Civil War (Bagley), 149–50
Soil Exhaustion as a Factor in the Agricultural History of Virginia and Maryland (Craven), 129
Soil Fertility and Permanent Agriculture (Hopkins), 44
soil jeremiads, 84, 92
Soil of the South, The (journal), 21
Soil Survey. *See* National Cooperative Soil Survey
"Soil Survey of Fairfield County, South Carolina," 54–56, 87
"Soil Survey of Stewart County, Georgia," 37, 56–61, 76, 87, 93, 172
Soils and Agriculture of the Southern States (Bennett), 58

Soils and Men (USDA), 90
Soils of Cuba (Bennett), 59
Soils of the United States (M. Whitney), 53, 54
Southern Homestead Act, 194
Southern Regions of the United States (H. Odum), 112
Southern River Swamp, The (Wharton), 188
Southerner Discovers the South, A (J. W. Daniels), 95–97
Squeeze, The (Higbee), 107
Steinbeck, John, 104
Stephenson, Lloyd W., 36, 37, 57, 173; *Preliminary Report on the Geology of the Coastal Plain of Georgia* (with Veatch), 33
Stewart, Daniel, 17
Stewart County, Ga., 12, 13, 14; activity of Geological Survey of Georgia in, 33, 36; activity of National Cooperative Soil Survey in, 37, 56–58, 87, 93; agricultural practices in, 170–72; Bennett visits, 56–58, 86; Bennett's invocation of, 59, 89–90; CCC camp in, 67, 75–76; early history of, 16–22; geology of, 172–77; visit by Cotting, 24–25; visit by Daniels, 95–97; visit by Flaherty, 103
Stewart-Webster Journal, 66, 67, 69–71, 77–78, 177
Stimson, Mrs. Henry, 72
Story of the Soil, The (Hopkins), 44–45
Stryker, Roy, 83
Sumter National Forest, 112, 198–99

Tallulah Gorge, 2
Talmadge, Eugene, 97
Taylor, John, 148, 161; *Arator*, 146
Teal, John, 188
Teal, Mildred, 188
tenancy, 151–56
Tenants of the Almighty (Raper), 155, 203
terracing, 127
Terrill, Helen Eliza, 70
Texas Agricultural Experiment Station, 85
Texas fever, 164–65

Third Southern Forestry Congress, 58–59
Thornthwaite, C. Warren, 93
To Hold This Soil (R. Lord), 92
Tobacco Road (Caldwell), 76
Toombs, Robert, 19
Tragedy of Lynching, The (Raper), 155
Treaty of Indian Springs, 17
Trimble, Stanley: Alcovy River swamps and, 188–92; correlation between slavery and erosion and, 146–47; land abandonment and, 135–36; Piedmont soil erosion and, 112, 141, 169–70; Piedmont stream sedimentation and, 122–23; mentioned, 114
Truman, Harry, 95
Tugwell, Rexford, 86, 91
Turner, Frederick Jackson, 129
Tuskegee Institute, 49, 200–202
Tuskegee National Forest, 200–202
Tuskegee Planned Land Use Demonstration (TPLUD), 200–202
Twain, Mark, 78

U-Haul, 2
Ultisols, 159–60, 162, 165
Uncle Remus, 202
Unified Farm Program, 203–4
uniformitarianism, 27, 29, 31
Union County, S.C., 112, 179, 198
Universal Soil Loss Equation (USLE), 166–70
U.S. Bureau of Outdoor Recreation, 81
U.S. Bureau of Soils, 37–38, 41–43, 45–46, 52, 54
U.S. Department of Agriculture (USDA), 32–33, 43, 45, 48; creation of, 28; grass research and, 163; soil conservation and, 85–86
—works: *The Land*, 104; *Soils and Men*, 90
U.S. Film Service, 103–4
U.S. Forest Service, 51, 121, 197, 202
U.S. Geological Survey (USGS), 31–34, 41, 52

Van Hise, Charles, 52
Vance, Rupert, 162, 163, 165; *The Human Geography of the South*, 150

Vanishing Lands (Jacks and Whyte), 94
Vasey, George, 163
Veatch, Otto, 36, 37, 57, 173; *Preliminary Report on the Geology of the Coastal Plain of Georgia* (with Stephenson), 33; *Second Report on the Clay Deposits of Georgia*, 33
Vertisols, 159
von Dongen, Helen, 104

Wallace, Henry, 86
Walter F. George, Lake, 13, 20
Warm Springs, Ga., 74
Washed Soils (Farmers' Bulletin), 49
Washington, Booker T., 49, 200
Weeks Act, 195, 197
Wharton, Charles H., 187–88; *The Southern River Swamp*, 188
Whitney, Eli, 119
Whitney, Milton, 38; biography, 40–41; feud with King and Hopkins, 41–44; National Cooperative Soil Survey and, 47–48, 50; Progressive conservation and, 52–54; retirement from Bureau of Soils, 59–60, 84; southern soil surveys and, 55–58
—works: *Soil and Civilization*, 54; *Soils of the United States*, 53, 54
Whyte, R. O., 94
Wilderness Act, 188
Williams, Bobby, 17
Wilmington Race Riot, 95
Wilson, James, 43
Wilson, Woodrow, 95
Wolcott, Marion Post, 83
Woodruff, Edwards, 205, 207
Works Progress Administration (WPA), 76; *Georgia: The WPA Guide to Its Towns and Countryside*, 76–77, 82

Yazoo Land Fraud, 130
Yellowstone National Park, 71, 101
Yosemite National Park, 71, 207–8
Young, Arthur, 134

ENVIRONMENTAL HISTORY AND THE AMERICAN SOUTH

Lynn A. Nelson, *Pharsalia: An Environmental Biography of a Southern Plantation, 1780–1880*

Jack E. Davis, *An Everglades Providence: Marjory Stoneman Douglas and the American Environmental Century*

Shepard Krech III, *Spirits of the Air: Birds and American Indians in the South*

Paul S. Sutter and Christopher J. Manganiello, eds., *Environmental History and the American South: A Reader*

Claire Strom, *Making Catfish Bait out of Government Boys: The Fight against Cattle Ticks and the Transformation of the Yeoman South*

Christine Keiner, *The Oyster Question: Scientists, Watermen, and the Maryland Chesapeake Bay since 1880*

Mark D. Hersey, *My Work Is That of Conservation: An Environmental Biography of George Washington Carver*

Kathryn Newfont, *Blue Ridge Commons: Environmental Activism and Forest History in Western North Carolina*

Albert G. Way, *Conserving Southern Longleaf: Herbert Stoddard and the Rise of Ecological Land Management*

Lisa M. Brady, *War upon the Land: Military Strategy and the Transformation of Southern Landscapes during the American Civil War*

Drew A. Swanson, *Remaking Wormsloe Plantation: The Environmental History of a Lowcountry Landscape*

Paul S. Sutter, *Let Us Now Praise Famous Gullies: Providence Canyon and the Soils of the South*

Monica R. Gisolfi, *The Takeover: Chicken Farming and the Roots of American Agribusiness*